Voice and Participation in Global Food Politics

As awareness of the commodification of food for profit at the expense of our health and the planet grows, this book foregrounds the communicative dimensions of resistance by food movements.

Voice and participation are argued by the author to be the means through which rural and urban communities can, and in many cases do, resist the capture of value by corporate actors and work to democratise their foodscapes. Her critical analysis of meaning-making under neo-liberalism suggests that agroecology, as a socially activating form of agriculture within a food sovereignty framework, provides an example of social learning relevant across rural/urban and North/South divides. Embracing indigenous knowledge, gender equity and postcolonial theory, this approach mobilises growers and eaters to contest the power structures that shape their food environments, and also to focus on social and economic justice within their communities, particularly in the context of climate change.

Participatory ecologies that incorporate these forms of social learning encourage the co-creation of inclusive foodscapes and politicise food justice. Such a positive framing of resistance through horizontal pedagogy, participation, communication and social learning processes contrasts with the vertical dissemination structure of the corporatised food regime and takes vital steps towards a more democratic food system.

Voice and Participation in Global Food Politics will be of interest to scholars of agri-food, transdisciplinary food studies and political economy of food systems. It will also be of relevance to NGOs and policymakers.

Alana Mann is Chair of the Department of Media and Communications within the Faculty of Arts and Social Sciences at the University of Sydney, Australia. Her research focuses on the engagement of citizens and non-state actors in activism and policy debates to inform the creation of just and sustainable food systems.

Routledge Studies in Food, Society and the Environment

The Real Cost of Cheap Food
(Second Edition)
Michael Carolan

Food Bank Nations
Poverty, Corporate Charity and the Right to Food
Graham Riches

Urban Food Systems Governance and Poverty in African Cities
Edited by Jane Battersby and Vanessa Watson

Risk and Food Safety in China and Japan
Theoretical Perspectives and Empirical Insights
Edited by Louis Augustin-Jean and Jean-Pierre Poulain

Localizing Global Food
Short Food Supply Chains as Responses to Agri-Food System Challenges
Edited by Sophia Skordili and Agni Kalfagianni

Seafood Supply Chains
Governance, power and regulation
Miriam Greenwood

Civil Society and Social Movements in Food System Governance
Edited by Peter Andrée, Jill K. Clark, Charles Z. Levkoe and Kristen Lowitt

Voice and Participation in Global Food Politics
Alana Mann

For further details please visit the series page on the Routledge website: www.routledge.com/books/series/RSFSE/

Voice and Participation in Global Food Politics

Alana Mann

Routledge
Taylor & Francis Group
LONDON AND NEW YORK

earthscan
from Routledge

First published 2019 by Routledge

2 Park Square, Milton Park, Abingdon, Oxon, OX14 4RN
605 Third Avenue, New York, NY 10017

Routledge is an imprint of the Taylor & Francis Group, an informa business

First issued in paperback 2020

British Library Cataloguing-in-Publication Data
A catalogue record for this book is available from the British Library

Library of Congress Cataloging-in-Publication Data
Names: Mann, Alana, author.
Title: Voice and participation in global food politics / Alana Mann.
Description: Abingdon, Oxon ; New York, NY : Routledge, 2019. | Series: Routledge studies in food, society and the environment | Includes bibliographical references and index.
Identifiers: LCCN 2018060190 | ISBN 9781138478343 (hardback) | ISBN 9781351068888 (ebk)
Subjects: LCSH: Food sovereignty. | Nutrition policy–Citizen participation. | Food industry and trade–Economic aspects. | Food industry and trade–Environmental aspects. | Sustainable agriculture. | Right to food.
Classification: LCC HD9000.5 .M3335 2019 | DDC 338.1/9–dc23
LC record available at https://lccn.loc.gov/2018060190

ISBN: 978-1-138-47834-3 (hbk)
ISBN: 978-0-367-73075-8 (pbk)

Typeset in Sabon
by Wearset Ltd, Boldon, Tyne and Wear

For Anne Dunn (1950–2012)
A beautiful voice.

Contents

Acknowledgements

This book is a response to the many voices that I have encountered in the course of my research on food movements. Families of the landless peoples' movement (MST) living the struggle for land justice in roadside encampments in Brazil; political economists raging against a failed development model in Delhi; farmers growing healthy food for their communities on permaculture farms in Cork, and *organipónicos* in Cuba; peasants uniting with the urban poor to reclaim land and livelihoods in Thailand; and scores of people doing food differently in kitchens, gardens, and streets all over the world.

Not all these voices or their stories can be included in any one book but they have inspired and sustained the writing of this one. As has the support of the University of Sydney, the Faculty of Arts and Social Sciences, and particularly the Sydney Environment Institute led by Iain McCalman, David Scholsberg, and Michelle St Anne. In a competitive academic climate where rankings rule it is a privilege to do and share research that has a real impact on the well-being of all species and the environment. We need to keep doing it.

Much of the thinking, talking, and writing was done during my sabbatical at Cornell University's Department of Development Sociology in the College of Agriculture and Life Sciences (CALS), and the Harvard Law School Food Law and Policy Clinic within the Center for Health Law and Policy Innovation (CHILPI). I thank all the colleagues, both here in Australia and abroad, who supported my visits.

Several diverse and equally enriching communities have been 'thought incubators' for the arguments I present in this book – the Australian Food Sovereignty Alliance, the Right to Food Coalition, the University of Sydney's Post-Truth Initiative, the Sydney Democracy Network, the Australia New Zealand Agri-food Network, the Symposium of Australian Gastronomy, the Australia New Zealand Communication Association (ANZCA), and the International Association of Media and Communication Researchers (IAMCR). All these welcoming spaces encourage the unruly transdisciplinarity, and conviviality, that comes with dwelling in the liminal space where food, people, and the planet intersect.

I give special thanks to the following friends and colleagues for ongoing conversations on the themes of this book, and much more: Benedetta Brevini, Nick Enfield, Bernardo Fernandes, Sumita Ghosh, Phil Hirsch, Eric Holt-Giménez, Donna Houston, Jesse Hsu, John Keane, Chris Mayes, Andrew McGregor, Phil McMichael, Eva Perroni, Elspeth Probyn, Emma Sharp, and Dave Mann.

Finally, enormous credit is owed to three of my former students who enabled me to deliver this manuscript in spite of their own commitments as working journalists: Drew Rooke for his thoughtful and professional editing; Justine Landis-Hanley for her dogged research on agribusiness mergers; and Alexander Tighe for assistance with manuscript preparation. I am so proud to have been a small part of the education of such forces of nature.

Chapter 1 contains extracts from Mann, A. (2018). Erasing Lines and Unsettling Truths: Lola Stayed Too Long. *The 'Anastasia' Project: Representing Heat through Performance*, *1*. Sydney Environment Institute.

Chapter 4 was originally published as Mann, A. (2015). Communication, Organisation, and Action: Theory-building for Social Movements. *Communication Research and Practice*, *1*(2), 159–173.

Chapter 5 reproduces and develops ideas published in Mann, A. (2018). Communication as Resistance in Food Politics. *The Political Economy of Communication*, 6(1), 36–58 and Mann, A. (2018) Education for Food Sovereignty as Transformative Ethical Practice, *Policy Futures in Education Special Issue: Eating in the Anthropocene: Learning the Practice and Ethics of Food Politics*, *15*(6): 744–766.

Chapter 6 includes the case study of Snowday previously published as 'Serving up food justice on the streets of New York' on the Sydney Environment Institute web site: http://sydney.edu.au/environment-institute/blog/serving-up-food-justice-on-the-streets-of-new-york/

Acronyms

AFNs	alternative food networks
ANAMURI	National Association of Rural and Indigenous Women (Asociación Nacional de Mujeres Rurales e Indigenas) (Chile)
ANAP	National Association of Agricultural Producers (Asociación Nacional de Agricultores Pequeños) (Cuba)
ANEC	National Business Association of Rural Producers and Traders (Asociación Nacional de Empresas Comercializadoras de Productores del Campo) (Mexico)
AOC	Acknowledgement of Country
ARC2020	Agricultural and Rural Convention 2020
ASCODE	Association of Central American Peasant Organisations for Cooperation and Development
ASPAN	Security and Prosperity Partnership of North America
ATTAC	Association for the Taxation of Financial Transactions and for Citizens' Action (Association pour la Taxation des Transactions financières et pour l'Action Citoyenne)
CAC	campesino a campesino (farmer-to-farmer)
CALD	culturally and linguistically diverse
CAP	Common Agricultural Policy of the European Union
CCO	constitutive role of communication in organisation
CET	Centre for Education and Technology (Chile)
CFAs	community food assessments
CFS	Committee on World Food Security
CIPP	Critical Indigenous Pedagogy of Place
CIW	Coalition of Immokalee Workers
CLADES	Latin American Consortium on Agroecology and Sustainable Development
CLOC	Latin American Coordination of Rural Organisations
CMP 8	the eighth session of the Conference of the Parties serving as the Meeting of the Parties to the Kyoto Protocol

COFCO	China National Cereals, Oils and Foodstuffs Corporation
CoP	Communities of Practice
COP 18	the eighteenth Conference of Parties (also known as the 2012 United Nations Climate Change Conference, or the Doha Climate Change Conference)
CSIRO	Commonwealth Scientific and Industrial Research Organisation
CSM	civil society mechanism
CSOs	civil society organisations
CSR	Corporate Social Responsibility
FAO	United Nations Food and Agriculture Organization
FoEE	Friends of the Earth Europe
FIAN International	Food First Information and Action Network
FPCs	Food Policy Councils
G8	Group of Eight
GHG	greenhouse gas
GM	genetically modified
GMFs	genetically modified foods
GMOs	genetically modified organisms
IAASTD	International Assessment of Agricultural Knowledge, Science and Technology for Development
IALA	Latin American Institute of Agroecology (Instituto Agroecologico LatinoAmericano)
IAPP	International Association for Public Participation
IFAP	International Federation of Agricultural Producers
IFPA-CRMB	Federal Institute of Pará-Rural Campus of Marabá
IMF	International Monetary Fund
IOM	International Organization for Migration
IPC	International Planning Committee for Food Sovereignty
IPCC	Intergovernmental Panel on Climate Change
IPES	International Panel of Experts on Sustainable Food Systems
LUSH	Landscaping for Urban Spaces and High-Rises (Singapore)
MACAC	Farmer to Farmer Agroecology Movement (Cuba)
MOICAM	Meso-American Movement of Indigenous and Peasant Farmers
MST	Landless Workers Movement (Movimento dos Trabalhadores Rurais Sem Terra) (Brazil)
NAFTA	North American Free Trade Agreement
NGOs	non-governmental organisations
PB	participatory budgeting
PCBs	polychlorinated biphenyls

PRONERA	Program for Education in Areas of Agrarian Reform (Programa Nacional de Educação na Reforma Agrária) (Brazil)
PUI	integral urban project (Proyectos Urbanos Integrales) (Medellín, Colombia)
RSC	Resource Consulting Services
SE	solidarity economies
SOCLA	Latin American Scientific Society of Agroecology
TANs	transnational advocacy networks
TRIPS	The Agreement on Trade-Related Aspects of Intellectual Property Rights
UA	urban agriculture
UAMPA	Union of Neighbourhood Associations of Porto Alegre
UFW	United Farm Workers
UNCCD	United Nations Convention to Combat Desertification
UNDESA	United Nations Department of Economic and Social Affairs
UNESCO	United Nations Educational, Scientific and Cultural Organization
UNFCCC	UN Framework Convention on Climate Change
UNHRC	United Nations Human Rights Council
UNORCA	National Union of Autonomous Regional Peasant Organisations of Mexico (Union Nacional de Organizaciones Regionales Campesinas Autonomas)
WTC	Welcome to Country
WTO	World Trade Organization

1 Introduction

We all eat. We all have our own foodscape, which includes 'the places and spaces where you acquire food, prepare food, talk about food, or generally gather some sort of meaning from food' (Mackendrick, 2014, p. 16). My foodscape, in Sydney's inner west, Australia, includes an organic cooperative, three supermarkets, a community garden, a weekly Saturday farmers' market, an abundance of restaurants and cafes, four hotels, and many fast-food restaurants – all within walking distance. I am fortunate – not only for having so many culinary options available to me but also for the simple fact that I can afford to eat. I am one of the 91.5 per cent of people in my city who are food secure. That looks like an impressive figure until you consider the flip side. Seventeen thousand people in my immediate proximity cannot always put food on the table for themselves and their families (City of Sydney, 2016). In 2018, food relief organisation Foodbank reported that more than four million Australians – 18 per cent of the population – had experienced food insecurity in the previous 12 months (Foodbank, 2018). Indigenous, culturally and linguistically diverse (CALD) and socially isolated people experience food insecurity at higher rates (Rosier, 2011). If you are of Australian Indigenous or Torres Strait Islander descent, you are five to six times more likely to be food insecure than other Australians (National Rural Health Alliance, 2016). In one of the most unaffordable cities in the world, food budgets often take a hit as rents and mortgages escalate while wages stay stagnant. Cities all over the world tell a similar story. Pockets of poverty and disadvantage can be identified through the diets and health of populations.

The scale of this problem is intimidating. *Or perhaps it is scale that is the problem.* Anna Lowenhaupt Tsing (2015, p. 38) argues that the expectation of scaling demands 'projects expand without changing their framing assumptions', therefore requiring project elements to be 'oblivious to the indeterminacies of encounter ... to allow smooth expansion' (ibid.). Scaling-up 'banishes diversity' when it is diversity that can generate change. Following Tsing, I suggest we turn our attention to the *nonscalable*, 'not only as objects for description but also as incitements to theory' (ibid.) in assessing the opportunities to improve our food environments. We purchase, prepare, share, and talk about food in our neighbourhoods. In a world where the near and the far are conflated by information technology and air travel, we still eat 'at home', in neighbourhoods where we are flung together by geography and economic interdependence. This is the scale that matters, that drives us to action, the basic unit of a polity which 'consists of people who live together, who are stuck with one another' (Young, 1996, p. 126). This is the first building block of wider food systems transformation as a collective project. We are in this together.

The illusion of choice

The negative elements of foodscapes are often blamed on consumers – we vote for unhealthy food with our forks. However, the rise of unhealthy foodscapes has more to do with 'deep-seated prejudices and the industrial food system's hunger for profit than the spatial distribution of food cravings' (Mackendrick, 2014, p. 17). Our food system is fundamentally undemocratic. It is dictated by Big Food – the agribusinesses, multinational food and beverage companies, and life science corporations that control supply chains. These actors have created a toxic food system that is killing us either through deprivation or through lifestyle-related diseases like diabetes and obesity. Unless we start to reclaim power over our food, we will have no say in how we produce and consume it in the future, and our bodies and minds will continue to suffer.

Even in affluent communities, the choice of food offered is an illusion. Power in food systems is now strongly embedded in the retail sector. People have to eat, and purchasing patterns suggest people prefer supermarket produce, *because this is what most people buy*. That is no surprise, given the lack of alternative foodways on offer to the average eater. In Australia, where I write this book, we have the second most concentrated grocery market in the world after New Zealand, and a cost price-squeeze is having a catastrophic impact on farmer livelihoods, particularly in times of drought. Governments on all sides of politics lack the political will to challenge Big Food, despite farmers insisting that, at the end of the day, they are unlikely to cover their costs, let alone make a profit. As more and more farmers leave the land, Australia's food security is increasingly under threat.

Eating healthy food produced in ways that do not damage the environment is essential to healing ourselves and the planet. This assumption is the foundation of this book. It is less about the problems of our food system which are detailed in growing abundance of academic and popular literature and more about talking, listening, practising and learning food together for potential social transformation. In this context, learning is both a process, and an outcome that promotes the emergence of assemblages of people, places and knowledges. It involves 'particular constituencies and discursive constructions, entails a range of inclusions and exclusions of people and epistemologies, and produces a means of going on through a set of guidelines, tactics or opportunities' (McFarlane, 2011a, p. 361). I argue that learning is essential to re-establishing connections with what we eat, and with the people who produce it, and is thereby central to the project of reclaiming control of our local food environments.

From this position, food becomes a mediator in learning processes. But if social learning is to bring about social change, we need democratic spaces and action centres for local and trans-local learning to engender transformative change across the entire food system.

A radical shift in consumers' mindset is also needed. We have all been 'deskilled and disempowered' (Booth & Coveney, 2015, p. 23) throughout the entire food chain – and the latter – disempowerment – is especially true of people experiencing disadvantage and malnutrition. We need to start re-understanding ourselves not merely as consumers but also as eaters or food citizens. And we need to include the voices of those marginalised, the 'all affected' (Fraser, 2007), if we are to realise new forms of food democracy (Lang, 2003) and take back control of our food.

Voice

Voice can be defined as, following Judith Butler (2005), a process of 'giving an account of oneself' in the form of a narrative. It is crucial in foodscapes, yet frequently absent. I share the concern of communication theorist Nick Couldry (2010, p. 1) that we are experiencing 'a contemporary crisis of voice' across not only political and economic but social and cultural domains. Couldry attributes this to the discourse of neo-liberalism which 'operates with a view of economic life that does not value voice and imposes that view of economic life onto politics, via a reductive view of politics as the implementing of market function' (ibid., p. 2) . In doing so it 'evacuates entirely' the role of the social in political regulation of economics.

To deny the capacity to possess and share one's narrative is to 'deny her potential for voice ... a basic dimension of human life' (ibid., p. 7). Socially grounded, voice requires resources in the form of language and status. It is a form of reflexive agency through which we 'disclose ourselves as subjects' (Arendt, 1958, p. 193) and make sense of our lives (Cavarero, 2000). As an embodied process of articulation, voice involves speaking and listening. It is therefore an 'act of attention that registers the uniqueness of the other's narrative' (Couldry, 2010, p. 9) which respects also the internal diversity or plurality in each voice. We all have many stories, embedded in multiple contexts.

Importantly, Couldry emphasises that voice is not about individualism, nor does it dismiss the value of collective forms of action. He emphasises that 'defending voice as a value simply means defending the potential of voices anywhere to matter' (ibid., p. 9).

> 'Voice' does more than value particular voices or acts of speaking; it values all human beings' ability to give an account of themselves; it values my and your status as 'narratable' selves.
>
> (Couldry, 2010, p. 13)

Accordingly, articulating one's voice is a form of resistance against an economic system which strives to silence voices, especially those that lack 'opportunity to compete as a commodity' (ibid.).

Given the domination of market logic over social life, where the economy comes 'pre-rationalised' or part of 'the given', Wendy Brown calls for a 'counter-rationality – a different figuration of human beings, citizenship, economic life, and the political' (Brown, 2003, para 42). Amartya Sen (2002, p. 10) insists we reclaim ethics in economic discourse through recognising human potential – i.e. 'the actual ability of [a] person to achieve those things she has reason to value' – while Nancy Fraser (2005, p. 75) argues for a reassessment of 'who is included in, and who is excluded from, the circle of those entitled to a just distribution and reciprocal recognition'. Étienne Balibar (2004, p. 114) demands a 'new civility' in the form of:

> ... a politics of politics [aimed] at creating, recreating, and conserving the set of conditions within which politics as a collective participation in public affairs is possible, or at least not made absolutely impossible.

In *Radical Hope* (2006) Jonathan Lear makes the point that whatever transformations our social, economic, political, and cultural institutions and structures undergo, they will lack value unless they are based on individuals' lived experience. Radical hope is grounded in valuing this experience, which can be translated into forms of 'counter-expertise' (Gilbert, 2008) that can lead to policies and politics for ordinary people.

Radical hope is evident throughout our foodscapes in the form of a mobilising civil society where pockets of inspiration and innovation are sites for doing food differently. Though some of these actors might not recognise themselves as part of it, there is a lot of discussion about whether a wider 'food movement' exists and, if so, how disparate groups unify to reclaim a food system that has effectively been corporatised. This emergent 'movement of movements' is comprised of energetic but often exhausted activists, advocates, practitioners and academics; it is fragmented more by time and capacity than ideology. Which raises some important questions: how do we converge in diversity to do the movement building necessary to bring about transformative change? Where is the common ground or common language needed to collectively combat the agents of Big Food who have created a system 'based on faith trust, reassurance, and unfortunately ignorance' (Booth & Coveney, 2015, p. 51)?

To reclaim our intimate experiences of food and repair the connections severed by its commodification, we can only start from our lived experience in the communities where we procure, eat and, if we are able, produce food. Our attachments to place include our food histories, and form our ways of seeing food. Learning can help us to recapture place-based ways. As Alethea Harper et al. (2009, p. 7) report in their evaluation of the development of Food Policy Councils in the United States, 'the power of informed, democratic convergence – especially when linked to the specific *places* where people live, work and eat – has an additional emergent quality: it can change the way we – and others – think'. Providing the structures and processes to enable this convergence of world views is not a straightforward task. The most food insecure are rarely invited, and even when they are, they often lack the capacity to engage in conversations about food policy. Capacity to participate, in terms of time, education, ability and will, comes before agency. Food insecurity is complex. It is often a symptom of severe and multiple disadvantage. Repairing relations between individuals in communities must come before reform. This healing is most difficult in countries which have not addressed histories of dispossession and violence such as Australia.

How did we get here?

According to Fritz Schumacher (2011, p. 8) 'one of the most fateful errors of our age is the belief that the problem of production has been solved'. He says this false belief is 'mainly due to our inability to recognise that the modern industrial system, with all its intellectual sophistication, consumes the very basis on which it has been erected' (ibid.). In the case of the food system, the relationship between capital and labour has been transformed to produce social exclusion on a scale we have never witnessed before. There are pockets of economic irrelevance in every country and city on earth, where

> areas that are non-valuable from the perspective of information capitalism, and that do not have significant political interest for the powers that be, are bypassed by flows of wealth and information, and ultimately deprived of the basic technological infrastructure that allows us to communicate, innovate, produce, consume, and even live, in today's world.
>
> (Castells, cited in Capra, 2002, p. 126)

Unlike ecological networks where no species or being is excluded and all contribute to the sustainability of the whole, in the world of wealth and power, large segments of the population are excluded or their value is limited to the supply of labour and raw materials. The neo-liberal trade regime, for example, has produced regions where local food production has been dismantled in favour of exports and imports, and transnational companies are free to relocate resource-extractive and polluting industries leading to environmental destruction. In these places, corporate power is exerted in a controlled, aggressive, and hierarchical manner, reducing the sovereign power of states.

Biotechnology provides a case in point. Technology is part of every human culture and predates modern-day science. It is not neutral; it has beneficial or catastrophic effects, depending on how it is applied, and who wields it. Genetic engineering has been promoted and adopted on a wide scale, with regulatory bodies and governments unwilling to exercise the precautionary principle under pressure from corporates. The overriding motivation for this is not advancing science or curing disease for the greater good, it is for financial gain. David Ehrenfeld, cited in Capra (2002, p. 163), argues that:

> like high input agriculture, genetic engineering is often justified as a humane technology, one that feeds more people with better food. Nothing could be further from the truth. With very few exceptions, the whole point of genetic engineering is to increase the sales of chemicals and bio-engineered products to dependent farmers.

The aspirations of life science companies such as Monsanto (which was acquired by the corporate agricultural giant Bayer in 2018) to 'consolidate the entire food chain' (Hodgson, 2018) underpin a corporatised food regime that concentrates knowledge and power in the hands of fewer and fewer merging mega-companies. This cedes control of food security to the profit-hungry, and represents a real threat to the democratic governance of food and agriculture policy. Framing market opportunities as moral imperatives, the agribusiness narrative is to 'feed the world' (Vidal, 2018) while making exorbitant profits at the expense of small-scale farmers and consumer health. This rhetoric, backed with scientific evidence produced by academic research funded by these very companies, leaves little room for alternative views. Corporate agents create and exercise discursive legitimacy in the public sphere by projecting and exploiting fears of a dystopian future based on two assumptions: first, that there is not enough food to feed the world and second, that genetic engineering is the only way to increase production. The rhetoric of scarcity is hollow; excess production is in fact the problem.

The food industry is a major contributor to overproduction, food insecurity and environmental degradation. This includes the production of up to a third of global greenhouse gas (GHG) emissions, when fertiliser manufacturing, food storage and packaging are taken into account. Yet Big Ag is committed to raising output, intensification of farming, mass processing, mass marketing, homogeneity of product, monocultures and a reliance on chemical and pharmaceutical solutions. The powerful US agribusiness lobby, for example, employs the rhetoric of scarcity to justify claims that America's farmers must double their production of grain and meat to meet the needs of a global population of nine million by 2050. In reality, the surplus, heavily-subsidised production of the US grain-livestock complex makes little contribution to ending global hunger and nutrition. This industrial monolith relies on the externalisation of costs, inequitable subsidies, and the (over)production of surpluses dumped as food aid while reducing biodiversity, and exhausting soils and water sources. The search for biomass alternatives to replace fossil-fuels drive land use changes which take more land out of agricultural production and place further stress on biodiversity.

Having a holistic view of the food system necessitates reconnecting the production of food with the downstream stages of the food chain. Most consumption happens in the urbanised regions of the world where the 'stuffed and starved', as Raj Patel (2009a calls them, co-exist. 'Food violence' (Eakin et al., 2010) in the form of hunger, obesity and diseases of malnutrition, disproportionately affect populations subject to chronic economic marginalisation, social exclusion and discrimination. These structural inequalities were highlighted in 2008 when food riots in over 40 countries were triggered by high food prices, lack of available food and reactions against government food policies. Signalling a 'critical stage in global neo-liberalisation' (Bohstedt, 2014, p. 16), the riots were driven by citizens joining in popular movements against perceived and actual breaches of the social contract by political leaders. At the time of writing, this malaise is manifested in the 'yellow vests' uprising in Europe. Violent protests in France triggered by fuel tax rises and living costs have led beleaguered President Emmanuel Macron to commit to an increase in the minimum wage, the removal of tax on overtime, and the elimination of surcharges on pensions (BBC News, 2018a). Whether this rhetoric will satisfy the discontented is dubious. 'We are fed up of hearing promises for politicians. The French don't believe in them any more', said Freddie Bouvier, a truck driver from Beauvais (BBC News, 2018b).

In times of austerity the problems inherent in the food system are by no means exclusive to the Global South yet the apparent abundance of food and comparatively low rates of food insecurity in more developed economies mask the unsustainability of how we produce and consume food as well as the rising number of hidden hungry. In the UK and Wales, for example, stagnating incomes and rising food prices since 2007 have made food 20 per cent less affordable for the lowest income decile than in the mid-2000s (Defra, 2014) and children are among those most affected (Lambie-Mumford & Green, 2017). As the Food Ethics Council (2017) reports, most indicators show the challenges facing the food system are getting worse. The current form of global capitalism is clearly ecologically and socially unsustainable, and therefore unviable. While we need more strict regulation, just business practices and fairly distributed technologies, they are not sufficient. We need deeper *systemic change*, particularly as capitalism's agricultural model approaches a critical juncture as climate change accelerates the collapse of food systems.

Food in a changing climate

Climate change is already having a dramatic effect on our food systems. In the future, the price of food is expected to rise dramatically while the quality and seasonal availability of produce will be reduced. Globalisation and urbanisation have increased homogeneity of food supplies with serious consequences for nutrition, and vulnerability. Following the failure of the Kyoto Protocol to unite countries in a concerted effort to set strict GHG emissions goals, adaptation emerged as 'the only viable option for furthering climate change policy' (Schipper, 2009, p. 364). Adaptation is defined by the IPCC as 'adjustment in natural or human systems in response to actual or expected climatic stimuli or their effects, which moderates harm or exploits beneficial opportunities' (IPCC, 2001). Adaptation planning should build the 'adaptive capacity' or the 'potential for actors within a system to respond to changes, and to create changes in that system' (Chapin et al., 2006, p. 16641). Countries need to reduce the GHG emissions from agriculture and adapt food systems to cope with growing global demand and competition for resources. Government science agencies, for example the CSIRO in Australia, maintain that climate change will 'force [Australian] farmers to adapt' (Clarke, 2013). Yet the dominant narratives promoted by these experts support unsustainable farming systems, characterised by specialisation, intensification and economic concentration. These systems are incapable of guaranteeing long-term food security (Lawrence et al., 2013; see also Howard, 2016).

Reconceptualising food systems' adaptation as a dynamic process that includes multiple social actors requires understanding that adaptation actions can be characterised in terms of intent (autonomous or planned) timing (reactive, concurrent or proactive) and direction/origin (top-down or bottom-up) (Fujisawa et al., 2015). It is widely recognised that adaptive capacity is based on social learning, knowledge exchange, empowerment and the bridging of social networks, which are multi-scalar (local to global). Knowledge diversity is necessary to solve complex problems (Folke, 2004), just as diversity within systems more generally 'is what allows them to increase their resilience and adaptability to new conditions' (De Angelis, 2017, p. 279). It requires us to recognise and respect the ordinary knowledge of farmers and communities who are witnessing the temporal and geographical dimensions of climate change in local conditions.

The sociocultural dimensions of climate adaptation policies involve food producers in specific regional networks who share adaptation innovations and generate discourses that circulate and intersect with wider conversations regarding climate resilient food systems. Strategies of adaptation are required to mitigate climate impacts and maintain the regional stability of food systems, which incorporate all activities involving the production, processing, transport and consumption of food, including its governance and sustainability. These strategies must facilitate citizen engagement in food systems transformations. Adaptation options involve many actors including farmers and consumers, and range from the application of agroecological farming methods to a shift to plant-based diets. Many of these bottom-up adaptation actions are being implemented within local food systems. Regenerative agricultural techniques that lift soil carbon levels, for example, are being shared throughout Australia via informal peer-to-peer networks and more formal structures, including local and regional farmers' associations, growers' cooperatives, non-government associations and interest groups. Further, these techniques are being informed by a growing recognition of the value of Aboriginal land management strategies in 'sustaining a complex mosaic of ecosystems continent-wide ... a form of proto-agriculture' (Massey, 2018 p. 25).

Here is important to note that adaptation discourses themselves tend to limit and depoliticise the forms of power that are produced and reproduced depending on one's ability 'to influence, profit from and find security' (Taylor, 2015, p. 8). In his analysis of agrarian livelihoods in Pakistan, India and Mongolia, Markus Taylor defines adaptation less as a 'valid analytical tool' than a 'politically constructed concept' that erases difference in highly unequal contexts. While climate change needs to be addressed on a global scale through coordinated and cooperative efforts to limit emissions and mitigate impacts, vulnerability will continue to be produced and reproduced in specific ways in different places unless we, as a species, acknowledge the relationships between local vulnerabilities and established structures of power and privilege. Failing to acknowledge this reduces adaptation to an 'abstract appeal to defend communities from external environmental disturbances and threats' (ibid.) that denies the complexity of sociocultural and economic difference. Accordingly, climate change is not a demon unleashed on hapless humanity but a clarion call to consider how the living conditions of individuals, already highly unequal, are masked by the urgency to adapt as a species.

For how does one adapt when one is already a victim of dispossession through the accumulation of others? Amitav Ghosh goes beyond the accumulation thesis to identify the role of colonialism, empire, and imperialism as the foundations of vulnerabilities across the globe. In *The Great Derangement: Climate Change and the Unthinkable* (2016) he marks out Asia, where the Bengal, Irrawaddy, Indus, and Mekong Deltas are submerging faster than oceans are rising while the activities that accelerate this process – dam-building and extraction of gas, oil and groundwater – continue apace. Half a billion lives are at risk in these regions, the sites of global cities established under colonial rule like Chennai, Mumbai, and Kolkata. In reading a World Bank report that Kolkata, his home city, is particularly vulnerable to damaging floods, Ghosh fears for his elderly mother's safety. She believes Ghosh is crazy to suggest that these wild scenarios, developed from impenetrable data sets, are grounds to leave the family home. And indeed, how do we leave the places where our memories and attachments give our lives structure and meaning, even in the face of terrible threats? Ghosh realises an important human truth in this exchange: 'my life is not guided by reason; it is ruled, rather, by the inertia of habitual motion' (Ghosh, 2016, p. 54). He resolves that, if we are all to adapt to the oncoming flood, decisions must be made collectively and implemented by international institutions.

Ghosh's reference to the vulnerability of global cities raises a vital concern regarding the precarious state of our food supplies in times of crisis, especially given the increasing urbanisation of the world. In 2010, 50 per cent of the total global population – up from 10 per cent in 1990 – lived in urban areas (UNDESA, 2010). Yet even though urban areas, which are defined as containing a minimum of 50,000 residents, take up just 3 per cent of the earth's surface, they are responsible for approximately 70 per cent of global energy-related carbon emissions (IPCC, 2014). And while they are centres of intensive resource consumption, cities are also home to the food insecure even in affluent countries such as Australia (Dixon et al., 2011; Mann, 2016, 2018a). Given that 66 per cent of the global population is expected to be urbanised by 2050, it is critical that cities incorporate foodways into their planning for development and expansion.

In some ways, this has already started happening. Many cities are taking heed of environmental and socio-economic risks which have exposed the fragility of local and regional food supplies, and have started developing 'resilient city' strategies (100 Resilient Cities, 2019). This discourse recognises that city and hinterland are interdependent and multi-scalar (Meerow et al., 2016), comprising social, ecological and technical systems that extend beyond city boundaries into, for example, sites of peri-urban agriculture. Resilience is an integrative construct that helps explain how people and communities achieve health and well-being. It is defined variously as 'the ability to adjust in the face of changing conditions' (Pickett et al., 2004, p. 373); 'the capacity to rebound from destruction' (Campanella, 2006, p. 141); and 'the degree to which cities tolerate alteration before reorganising around a new set of structures and processes' (Alberti et al., 2003, p. 1170). Resilience and food security depend on 'maintaining production of sufficient and nutritious food in the face of chronic and acute environmental perturbations' (Bullock et al., 2017, pp. 880–884).

Optimistically, these demands are a potential bridge between the rural producers and the urban food justice movement. And while cities are 'machines for learning' (McFarlane, 2011a, p. 362) – 'spaces of encounter and rapid change, of concentrations of political, economic and cultural resources and of often perplexing unknowability' – rural spaces, and the liminal spaces in-between, are similarly labs for experimentation and innovation. In re-visioning our food environments, rural and urban entanglements are sites for *sharing* learning and joining in 'social revolution', defined by Massimo De Angelis (2017, p. 270) as

> a process of finding solutions to the problems that capital systems cannot solve, because it [*sic*] has created them, and the rest of us have an urgent need to address them: social justice, a dignified life for all, climate change, environmental disaster.

Bringing politics to the table

Food is a highly tangible indicator of social inequality. As Alexandria Ocasio-Cortez, the youngest woman ever elected to US Congress stated in an interview with *bon appétit* magazine in November 2018,

> The food industry is the nexus of almost all the major forces in our politics today … linked with climate change and ethics … minimum wage fights, of immigration law, of criminal justice reform, of health care debates, of education.
>
> (Cited in Cadigan, 2018)

Yet in affluent economies, we tend to depoliticise issues of food politics by focusing on consumption behaviours at the expense of organising. We need to evolve from consumers to food citizens and reclaim our agency (Friedland, 2008). This form of deeper democracy is possible within solidarity economy approaches where communities work collectively to provide employment, healthy environments, affordable healthy food and collective ownership. Truly participative models go beyond formal institutions of democracy and the mass-mediated public sphere to include and empower society's most marginalised. Diverse partnership and governance models can create enabling environments for the co-creation of healthier and more just food systems on scales ranging from neighbourhoods to cities and nations. They empower and engage local communities in acts of movement building that resonate trans-locally. Participatory cultures and forums for community members to come together and address common concerns about food policy on a local level, from food policy councils to more radical commons ecologies where individuals experience personal processes of emancipation through learning and empowerment, are essential if we are to help communities help themselves. In these spaces, the common language between the fragmented food movements can be discovered and deployed to collectively combat the rhetoric of a food system that we have effectively outsourced to profit-making companies.

The 'multi-scalar systems of social action' that 'seek an alternative path from the dominant ones [and are] able to reproduce at greater scale through networking and coordination' include social movements, cooperatives, self-managed public spaces and policy instruments such as participatory budgeting. These are all 'social systems at different scales of action within which resources are shared, and in which a community defines the terms of sharing, often through forms of horizontal social relations founded on participatory and inclusive democracy' (De Angelis, 2017, p. 270). La Vía Campesina (LVC), 'the peasant way', is a striking example. Claiming to represent 200 million small-scale producers, fishers, and landless workers in over 70 countries, this movement links far-flung rural constituents supporting fair prices for farmers; ethically and environmentally responsible production methods; appropriate regulation and trade policy; the breaking down of agribusiness monopolies; workers' health and safety; and consumer access to healthy, nutritious, and culturally appropriate food. LVC is fundamentally focused on promoting the concept of 'food sovereignty' based on principles of agroecological production. Pioneered by indigenous, peasant and rural proletarian groups in Latin America (Meek, 2014), 'food sovereignty' is understood as 'the right of peoples to healthy and culturally appropriate food produced through ecologically sound and sustainable methods, and their right to define their own food and agriculture systems' (Declaration of Nyéléni, 2007). It is explicitly an anti-capitalist concept, privileging social reproduction over capital accumulation. As an 'active anti-systemic model' (McMichael, 2009) with an 'anti-hegemonic agenda' (Wolford cited in Hall, 2014) it represents the most radical response to the corporate food regime. Far from a simple form of agrarian populism, its legitimacy is widely recognised within global agencies including the United Nations Food and Agriculture Organization (FAO) where it plays a major role in the Civil Society Mechanism (CSM) of the World Committee for Food Security (WCFS).

Food sovereignty extends beyond farmers' interests towards a democratic political project embracing themes of diversity, inclusivity and social justice. Contesting ontologies, identities and the operation of power, the farmers fighting for it demand inclusive political spaces where they expose the unsustainability of a food system dependent on liberalised markets, the increasing power and reach of transnational corporations, advances in biotechnical solutions, exploitation of labour and the encroachment of regulations in both public and private spheres. As well as serving as a model of organisation and resistance through communication LVC is adept at framing solutions. Openly rejecting the notion of food as a commodity it adopts a rights-based approach that asserts the right to food is indivisible from other human rights including gender and racial equality. Supporting collective and social rights, LVC creates spaces for the participation of those most affected by hunger. This politicised notion of citizenship expands to include food along with rights to housing, health and education. As meaningful participation in food democracy, it involves individuals becoming knowledgeable, sharing ideas, and developing efficacy with respect to the food system, underpinned by an orientation to the greater good. The durability, resilience and progress of the movement demonstrates that while hunger may drive violence and conflict, it can also drive new forms of collaboration and cooperation (Mann, 2017a).

We can learn a lot from the rural proletarian social movements that comprise LVC about how communities can be granted agency to participate in the solving of the problems that affect them most. This includes recognition of other standpoints and ways of knowing, coupled with the sharing of knowledge horizontally, both locally and in transnational circuits. For instance, food sovereignty recognises that indigenous peoples worldwide have suffered from European colonisation leading to the removal or alteration of traditional lands that produced a variety of traditional foods, and the obliteration of their foodways. Environmental degradation, neo-liberal trade agendas, lack of access to land, the breakdown of tribal social structures and socio-economic marginalisation are among the barriers to healthy and culturally adapted indigenous foods.

The discourse of food sovereignty privileges indigenous views, knowledge and practices in biodiversity conservation and recognises that indigenous territories include the world's remaining areas of highest biodiversity. Aboriginal conceptions of food sovereignty emphasise food as sacred, reflect deep connections/kinship with the environment and rely on intergenerational transmission of food-related knowledge. As Bruce Pascoe tells us in *Dark Emu: Black Seeds: Agriculture or Accident?* (2014) we ignore the knowledge of the land and growing food accumulated by Indigenous Australians over many thousands of years at our peril. Aboriginal people developed highly complex social mechanisms, cosmologies, religious and kinship systems resulting in sophisticated and sustainable ecosystem management by combining 'ecological literacy combined with lack of ego' or '[getting] out of the way of mother nature' (Massey, 2018, p. 378). Based on recognition that nature contains self-organising systems and drawing on collective memory in the form of 'songlines' that trace the 'ecological spines' of the country, First Nations peoples practised regenerative and sustainable land management techniques such as firestick farming and terracing. Our current landscapes are evidence of their active long-term assistance to the land (Massey, 2018; see also Gammage, 2012; Pascoe, 2014).

This reminds us that the European settler experience of food is built on dispossession. If it is indeed true, as environmental advocate Gus Speth asserts, that 'today's problems cannot be solved with today's mind' (cited in Massey, 2018, p. 418) then it is time to reflect on how we acquire knowledge and make decisions, and what knowledge we have dismissed through our arrogance. Barriers to constructive solutions to problems include deeply embedded world views that are siloed into specific domains, disciplines and cultures. Valerie Brown (2007) makes the case for synthesis-based thinking that challenges dominant Western values of competition, individualism and hierarchy. These are the same values that drive our capitalist economy, the source of the ills of the food system. Collective thought and action requires redesigning the social structures and framework that support the divisions between our different ways of knowing the world, whether these be based on science, society, traditional knowledge and/ or lived experience.

As an initial step we must resituate our foodways within a living system in our thinking. Eating food, among the most intimate and organic of activities, has taken on an instrumental function for too many of us. Food as fuel has replaced food as culture; food as a commodity has replaced food as identity. Our sense of contingency in food systems has disappeared. Instead of responsibly creating and recreating environments for the future, we focus on 'the short-term accumulation, expropriation and exhaustion' of natural resources as if 'nothing new or too catastrophic will occur in the future, and, hence, the way that things are today will continue to be the way of things in the future' (Guardiola-Rivera, 2011, pp. 29–30). A holistic view of the food system involves recapturing a sense of equilibrium and contingency represented in the symbols of ancient cultures such as the Australian Indigenous and Amerindians

> dots, zigzags, scales, spirals ... observations about the contingency and fragility that characterises everything that exists in creation: floods and droughts, rainy season and dry seasons, disappearing and reappearing stars in the sky, growth and decay, death and rebirth, the rise and fall of rulers and cities.
>
> (Ibid., p. 29)

These traditions, though ancient, are not irrelevant. In fact, only through these deeply embedded notions of mutuality can we repair and rebuild communities and create deep, meaningful connections on the basis of fair and equitable access to and exchange of what should be a common good – food. Our project to re-diversify the 'monocultures of the mind' (Shiva, 1993) that have led to the dominance of conventional models of agriculture means overcoming the instrumental and economic rationality embedded in industrial food production. Transformative approaches to thinking and learning are embedded in the praxis of many rural social movements. Drawing on the previous struggles of members, their knowledge and experience, these movements build programmes *within* and *from* communities, not *for* them (O'Cádiz et al., 1998). Where local and traditional knowledges have been historically excluded from land management and policy-making, LVC includes them through *diálogo de saberes* – dialogue among different knowledges and ways of knowing. This practice is 'key to the durability of the LVC constellation ... as organisations take mutual inspiration from the experiences and visions of others' (Martínez-Torres & Rosset, 2014, p. 980).

The movement towards 'community self-reliance and an economy rooted in human solidarity rather than amoral competition' (Boggs, 2012, p. 170) is more familiar to Asian and Latin American countries than Australia and the United States. Yet seasoned activists such as Grace Boggs

(2012) express hope with the emergence of cooperatives, mainly credit unions, in which 130 million Americans participate and the rise of employee-owned companies and community development corporations. These experiments in collective ownership and management are not always successful but they are a necessity given the economic insecurity of millions with housing unaffordability, unemployment, and a rapidly ageing population. Precarity, a condition of living with little or no predictability or security is everywhere as each person 'must devote her life to sustaining competition, in conditions of every-increasing resource constraints and corrupted goals, which reproduce the same collective problems' (De Angelis, 2017, p. 3).

Seeking synergies

This book is a small contribution to the growing call for a 'radical transformation of our world' based on people coming together into communities to develop alternatives to the logic of capitalism and 'multiply them and interconnect them' (De Angelis, 2017, p. 11). Its scope is conceptually and geopolitically broad, embracing examples of the democratisation of food systems across rural and urban landscapes internationally. It aims to demonstrate how communities are doing food differently, in a revolutionary way. Some of the most successful initiatives are small-scale and community based. They are being shared trans-locally. They are the seeds of transformative change in the food system. They embrace collective thinking and action from different cultures and demonstrate that much can be learnt from truly collaborative engagement with other ways of knowing the world.

Crisis does provide opportunities for the presentation of ideological alternatives (Wade, 2009; Munck, 2010; Veltmeyer, 2010), some of which are discussed in this book. Though differing in their specificities, all these alternatives stress the need for new ways of producing, procuring, and consuming food; operationalising them in various contexts is the challenge. An eco-literate and socially just approach to food systems planning starts with a fundamental question – whose knowledge counts in debates about food and agriculture?

If the crisis of our food system is fundamentally a crisis of democracy, we need to regain agency and control over the way our food is grown, distributed, and processed. But what does a food system that creates solutions from the bottom up look like? M. Jahi Chappell and Mindi Leigh Schneider (2016) propose a three-legged stool of agroecology, food sovereignty and food justice to replace the dominate tripartite of social, economic and environmental sustainability. This approach, they argue, acknowledges that agri-food systems are significant drivers of damage to human and planetary health, and promotes agroecology as regenerative practice. It also transcends disciplinary boundaries, as well as the dichotomy between nature and society as two distinct systems, and frames the agri-food system as a problem for sustainability but also as a solution. Food justice is tied to degrowth (*décroissance* in France, where the movement originated) and addresses racial, gendered, and class-based inequality. Alkon and Agyeman (2011, p. 8) define it as 'mirroring' the two key concerns of food sovereignty (as defined earlier) and food access, which refers to the ability to produce and consume healthy food.

Food justice also has strong links to environmental justice in the way food is produced, consumed and disposed of. Heynen et al. (2012, p. 306) suggest, however, that 'for *food justice* to have intellectual and political value, it must both take advantage of the robust history of food politics and then move these politics forward toward more emancipatory goals'. They propose that urban agriculture can link *community food security* to *food sovereignty* in order to realise *food justice*. Chappell and Schneider (2016) present agroecology as a 'more robust accompaniment' to sovereignty and justice that links rural and urban environments. I follow this line of thought with a particular emphasis on the power of agroecology as a methodology of emancipation to politically mobilise not only rural producers (Perfecto et al., 2009) but urban growers and eaters. Emancipation, as the '*process of liberation*, of being set free' is about individuals; it 'does not allow for prescriptions or models; it is a process that is always unfinished and must be experienced individually ... it is not an objective, it is a way of life' (Zibiechi, 2012, cited in De Angelis, 2017, pp. 357–358).

All three legs of the stool are necessary to ensure food security without devastating the natural environment. The integrative approach to agroecology is 'transdisciplinary, participatory and action-oriented', and to be transformative must include a critique of the political economy of the food system as it exists today (Méndez et al., 2013, p. 3; see also de Schutter, 2017). For LVC, agroecology and food sovereignty are inextricably linked on the basis that food sovereignty values traditional farmer and community knowledge which is incorporated into agroecological practice. It represents 'a way of life that recovers all we have lost, a connection with ancient knowledge' (La Vía Campesina, 2013a) that includes modes of exchange and trade outside the neo-liberal market model. It opens space for solidarity 'economies and transition movements' (Sage, 2014) based on respect and reciprocity. These learnings are vital in urban contexts where the needs of those who do not produce food are still impacted by undemocratic agri-food systems through poor access and availability to healthy and affordable food. We are all impacted by the 'bads' of the food system. Everybody eats.

Basing my thesis on research undertaken primarily in Latin America, Europe, and Australia I propose that we look trans-locally for models of the transformative thinking required to radically change our food system. For example, to Brazil where models of participatory budgeting have improved the delivery of social services. To Cuba where farmer-to-farmer knowledge sharing has accelerated agroecological land management. To London where austerity programmes have driven local boroughs to adopt radical new approaches to healing marginalised communities. These activities aim to address social isolation, and many involve food as a mediator in social relations. These activities are focused on repairing and rebuilding social capital through providing inclusive spaces for people to participate in daily life.

A word on methodology

In focusing on the values of voice, participation, and learning, I adopt a transdisciplinary approach grounded in the social sciences. Admittedly social science is, in the words of Raewyn Connell (2007, p. vii), 'ambiguously democratic'. It is generally based on the views of dead white men in Europe and North America. Connell's recognition that while situating knowledge of society in experiences outside the 'global metropole' of Western civilisation is a 'fragile project' it is essential, as 'only knowledge produced on a planetary scale is adequate to support the self-understanding of societies now being forcibly reshaped on a planetary scale'. Where Oscar Guardiola-Rivera (2011) asks 'what if Latin America ruled the world?' I ask what can Latin America teach us about doing food differently? Of course, food systems change is not proceeding unchallenged in the countries I discuss but it does have lessons to teach us about the power of forces of mobilisation, collective action and pedagogies of empowerment to bring about the radical change needed to transform our food environments. Guardiola-Rivera (2011, p. 9) refers to the uniquely Latino combination of

> older languages and practices of reciprocal recognition with the ethics of fraternity of associations and the modern language of public recognition through law, and also with newer understandings of equality and freedom as self-determination, sovereignty over resources and control over their own political and economic paths.

For these and other reasons, Latin American cities have evolved as a 'prime testing ground' (McGuirk, 2015, p. 23) for radical ideas in urban planning and governance, themes that are explored in Chapter 6 of this book.

In this research I challenge the traditional reference points of global geopolitics – east–west, west–rest. They are inadequate to account for the fresh and dynamic positionalities of citizens. Like many Western scholars, I grew up learning about the exclusive diffusion of institutional reform from the north to south, and the dominance of the north in generating these. South–south connections are found in the work of many of the University of Sydney colleagues I have had the pleasure of working with, including Irene Strodthoff (2014), Fernanda Peñaloza, and Sarah Walsh

(2019). Recognising that researcher positionality has influence on the research process, I explicitly declare myself as an 'outsider' to many of the cultures I have studied (Hammersley, 1993; Herod, 1999). Undoubtedly my personal world view is coloured by my values, race, and gender (Sikes, 2004), coupled with more subjective elements, such as my history and experience (Chiseri-Strater, 1996). This reflexivity leads me to self-assess my location in relation to the subject, the participants and the research process, including the design and interpretation of my findings (Greenbank, 2003; Savin-Baden & Howell Major, 2012). My mixed method approach is pragmatic, determined by my previous doctoral research on Latin and Iberian agrarian movements (Mann, 2014a); my professional background as an educator in secondary schools in my early career, and now university; and my personal interest in pedagogies for empowerment. I am keenly aware of the nature of my position as an educated, Australian woman of European descent. However, as I have discovered in my research in countries including Mexico, Brazil, Chile, and Cuba, this does not necessarily predispose me to a colonial or settler perspective. My primary aim is that my research is of value to those who make it possible – the agrarian, colonised communities who promote alternative, socially just ways of producing and providing food as the means to achieving *buen vivir*, or collective well-being:

> ... a system of knowledge and living based on the communion of humans and nature and on the spatial-temporal-harmonious totality of existence. That is, on the necessary interrelation of beings, knowledges, logics and rationalities of thought, action, existence and living.
>
> (Salazar, 2015)

Organisation of the book

Following this Introduction, Chapter 2 details the methods involved in the corporate capture of not only voice but value and knowledge practices. I introduce food regime theory as a lens through which we can observe the operation of power via 'legitimizing ideologies' (McMichael, 2013, p. 8) such as free trade, food aid, privatisation, and biotechnology. I focus on the latter in interrogating how agribusiness mega-corporation Bayer-Monsanto discursively legitimises its imperative to feed the world at the expense of the health of people and the planet.

Chapter 3 engages with wider questions of voice, participation and learning in food systems, presenting the theoretical frameworks that underpin the succeeding chapters. Starting with repair and recovery as key elements of social inclusion, I interrogate various models of civic engagement as ways to include those most affected by the 'bads' of the food system in the co-creation of their food environments, before moving on to the concept of emancipatory, social learning as a key to empowerment and collective action.

The theme of collective action is followed in Chapter 4 with a case study of convergence, in diversity, through communication. Applying constitutive role of communication in organisation (CCO) theory (McPhee & Zaug, 2000; Cooren et al., 2011) I analyse how LVC has been so successful as a transnational coalition in an effort to better understand what the wider food movements can learn in terms of governance and methods, while problematising the challenges of being truly representative.

Central to LVC's success has been the articulation of the mobilising frame of food sovereignty, and its inextricable connection to agroecology as an alternative to conventional agriculture. Chapter 5 discusses agroecology as resistive epistemology. Case studies of Schools of Agroecology in Cuba, Chile, and Brazil reveal how resistive epistemologies are a source of empowerment and agency at the grassroots. In educating and empowering the marginalised to become political agents, agroecology is revealed to be not only a farming method but also a framing device that has emerged from, and incorporates, social movement dialogues.

Chapter 6 presents different expressions of countermovement against the industrial food system, with a specific focus on urban contexts. I discuss the importance of urban agriculture, food-centred design of cities, and governance models that include public-private partnerships, food policy councils, and innovative technology platforms. I present international examples of learning networks that place food at the centre of projects to address social isolation, food insecurity, waste and provide fair returns to producers. Many of these cases demonstrate trans-localism as a way of scaling across rather than up in co-creating transformative approaches to food systems.

Trans-localism, a site for doing, performing, experimenting, practising and sharing things differently, for consolidating 'networked relationalities' between food production and consumption (Sonnino, 2017, p. 5), requires an inclusive and vibrant public sphere. People need to be supported and motivated to participate. This theme is revisited in the concluding chapter, Chapter 7, where I review the communicative strategies, tools and institutions we need to build better food environments through voice, participation, and learning.

Chasing utopias

> Utopia is on the horizon: When I walk two steps, it takes two steps back. I walk ten steps, and it is ten steps further away. What is Utopia for? It is for this, for walking.
>
> (Galeano, 1971)

Earlier utopian versions of civil society theory involve social movements as connectors between public opinion and public policy. Utopian visions should not be dismissed as populism but understood as motivations for people to challenge the status quo, in the first place, and they can create the political will for radical, structural change that reduces oppression. *Real* utopias are based on ideals of the 'potentials of humanity' and, as such, provide 'pragmatically accessible way stations ... that can inform our practical tasks of muddling through in a world of imperfect conditions for social change' (Fung & Wright, 2003, p. vii). Food utopias are necessarily 'hope-filled' (Stock et al., 2015, p. 3) as they offer a better way of living. More than just a mix of alternative food initiatives on a local level that offer innovative while limited strategies for change, they can *re-politicise* our thinking about how food systems are created.

I do not present the models in this book as examples of utopia. Drawing on the lessons of countermovements such as LVC is to recognise that their world views are embedded in traditional land and ecological systems that oppose the commodification and marketisation of not just food, but productive resources and human lives. These views mirror the traditions of many other cultures and communities concerned about the deeper aspects of social justice. They provide us with a way of mapping the influences of global over the local and demonstrate how we can share knowledge trans-locally to help advance far-reaching social and political change.

We need utopias *to get somewhere*. In this respect, rather than representing 'a better world like it is a given, unchanging location' any food utopia becomes 'the necessary act of imagining a better world, which involves processes for learning, adapting, and changing in response to growing knowledge and shifting circumstances' (Chappell, 2018, p. 26). It calls for a deeper democracy and 'visionary organising', a time of convergence together 'as inventors and discoverers committed to creating ideas and practice, vision and projects to help heal civilisation' (Boggs, 2012, p. xxiii).

The time for repairing, reclaiming, and re-visioning our food systems has arrived.

2 The capture of voice and value

Corporate interests have reached their zenith under neo-liberalism as market function is adopted as the dominant priority for social organisation. Promoting the market as the solution to social and economic problems and providing infrastructure to distribute goods, multinational corporations serve as vehicles for neo-liberal ideologies. Their domination of, reliance on, and subservience to the market as a form of 'hegemonic rationality' reduces the complexity of social world to 'spaces of potential competition that need to be organised *as* markets, blocking other narratives from view' (Couldry, 2010, p. 6). As illustrated in this chapter, obstructing the voices of actors who do not participate in the market eliminates the possibility of those actors to control their own accounts of themselves. This deep denial of voice is an oppression described aptly by W.B. Dubois as a 'sense of always looking at oneself through the eyes of others' (cited in Couldry, ibid., p. 9) – a condition familiar to those who experience marginalisation (and categorisation) on the basis of race, class, gender and socio-economic status. This 'voice-denying rationality' through the market-driven organisation of social life involves, at its most extreme, the organisation of resources 'on the explicit basis that some individuals' voice and life [has] no value' (ibid., p. 10). The impact of this is witnessed in the expansion of markets into the Global South, the devastation of local livelihoods and the subsequent forced migration of poor and landless workers across borders. La Vía Campesina (LVC) observes that the global movement of food leads to the global movement of people. Phil McMichael (2013) extends this correlation to the movement of *money* as a primary driver of migration as the financialisation of food and land dispossesses the rural poor on every continent.

Central to this discussion is the capture of value in our food systems. Food, along with the human labour and resources required to produce it, has been stripped of its value in capitalist agriculture. The expression 'dirt cheap' belies the fact that arable soil is at a premium – contributing globally US$16.5 trillion annually in ecosystem services yet disappearing at a rate of 24 billion tons each year through erosion (FAO and UNCCD cited in Coleman, 2017, p. 175). Our failure to replenish the soil at the rate at which it is removed by urbanisation, desertification and industrial agriculture is leading some scholars to promote the idea of *soil as a currency*. Are we so incapable of valuing such an important resource that we have to monetise it in the manner of water, forests and carbon? This is our quandary in a system that 'values only what it can count, and it can count only dollars' (Patel & Moore, 2017, p. 21).

It is profit-seeking that explains why food is cheap. Food is meant to be wasted rather than feed people in a system when its commodification 'denies its non-economic attributes in favour of its tradable features, namely durability, external beauty and the standardisation of naturally-diverse food products' (Vivero-Pol, 2017, p. 3). Commodification leads to food insecurity as it privileges cheap calories over nutrition and scales up production to create surpluses. Market concentration places power in firms while it disempowers growers and eaters. Food regime theory, in this chapter, provides a useful lens through which to view how these macro-economic forces impact on diets, health and livelihoods in Mexico, an extreme case. It explains how the dynamics of the corporate food regime (McMichael, 2009) have enabled multinationals such as Bayer-Monsanto, also profiled here, to profit from the commodification of food and nature.

Life under the corporate food regime

Food regime theory helps us to understand how 'the food chain binds and transforms the world's different cultures via commodification' (McMichael, 2013, p. 1). Food regimes emerge from power contests between social movements and powerful entities including corporations and governments who set rules, relationships and practices that 'come to seem natural … when the regime works really well, the consequences of actions are predictable, and it appears to work without rules' (Friedmann, 2005, cited in McMichael, ibid., p. 10). This calls to mind Eric Holt-Giménez's observation in *A Foodies' Guide to Capitalism* (2017) that the food system is not 'broken' – it works very well for the companies who profit from it.

A food regime is defined as a 'rule-governed structure of production and consumption of food on a world scale' (Friedmann 1993 in McMichael, 2009). Each regime establishes 'particular conditions' through stable relationships between elements of the supply chain and world prices dictating the production, circulation, and consumption of cheap food (McMichael, 2013, p. 11). Global markets are structured specifically for cheap food, the production of which exploits cheap land and labour for exports from the South and dumping from the North. In this way food regime theory provides a 'comparative-historical lens on the political and ecological relations of modern capitalism writ large' (McMichael, 2009, p. 142).

The first food regime (1870–1930) was based on colonial tropical imports including grains and livestock produced in settler colonies through monocultural agriculture, a simplification that has reached crisis point with biodiverse and sustainable agriculture increasingly under threat. In the second food regime (1950–1970s) surplus flows of food were sent to post-colonial states as food aid. This created the development model, an enduring power structure that creates needs and then proposes to satisfy those needs – a 'subtle strategy' that enables interventions by nation-states, NGOs and multinational corporations in the guise of benevolence (Giraldo & Rosset, 2017; see also Escobar, 2011). Green Revolution technologies accelerated yields while transnational linkages between national farm sectors grew into global supply chains and created a new international division of labour in agriculture. In Latin America, landscapes such as the Cerrado, Amazon and Pampa have undergone a 'brutal transformation' at the hands of agribusiness that began with the green deserts of sugar plantations under slavery in the seventeenth and eighteenth centuries and is now seeing 'endless fields' of soya beans, maize and intensive livestock farms (Rundgren, 2016, p. 104). Under these conditions commodity markets grow and, through financial speculation, contribute to food price hikes such as those experienced in 2007, 2008 and 2011. Overproduction is manifested in exported surpluses that propel the expansion of agribusiness transnationals (Weis, 2007; Guthman, 2011).

The third food regime (late 1980s–now) is characterised by McMichael (2008) as a deepening and expansion of the co-existing, second regime to emerging economies such as China and Brazil, accompanied by a power shift to the retail sector via a 'supermarket revolution', the emergence of global food/fuel complex and the end of the liberalised trade regime with the decline of the influence of the World Trade Organization (WTO). As these conditions extend across farming regions and the entire food chain, their costs are externalised and manifested in climate change, biodiversity loss and the poisoning of water and soil. Accordingly, 'the trajectory of the corporate [third] food regime is constituted through resistances: both protective (e.g. environmentalism) and proactive, where "food sovereignty" posits an alternative global moral economy' (McMichael, 2005, p. 286). The moral economy concept, which was first applied to popular protests against capitalist expansion in the eighteenth century (Thompson, 1971) and was later applied to forms of tenure enforced by the French and Dutch colonial powers (Scott, 1985) is today the basis of the Brazilian Landless Workers Movement's (MST) argument that productive peasant farms are more valuable than the cash crops harvested from monocultures such as sugar cane, for example. Food sovereignty captures the contradictions between a '"world agriculture" (food from nowhere) and a place-based form of agroecology (food from somewhere)' (McMichael, 2009, p. 147). Agroecology provides the 'moral economic basis' (Meek, 2014) of an alternative to the industrial food system.

Here McMichael questions – what is alternative? He argues that industrial agriculture should be considered the alternative, given the majority of the world's 'extant food cultures' follow a 'healthy logic of reproduction of social and ecological relations, as opposed to the degrading and disabling force of capitalist agriculture's dynamic of under-reproduction of social labour and ecosystems' (McMichael, 2013, p. 131). The food sovereignty movement aims to restore sustainable capacities through 'broadening and recovering values that reconstruct diversity and supersede the homogeneity of the exchange value regime' (ibid.). By exchange value McMichael is referring to the bio-economy as the highest stage of commodification, whereby crops such as corn, soy and palm oil become 'fungible investments' virtually interchangeable as feed, fuel or fertiliser. In these transformations exchange value 'erases' actual value in the 'ultimate fetishisation of agriculture, as an input-output process geared to indiscriminate production of commodities for profit' (ibid., p. 132). To LVC this is 'agriculture without farmers' (cited in McMichael, ibid.).

The transformation of public goods, the 'cornerstone of human rights', is now accelerated by the emerging processes of dematerialisation, digitalisation and financialisation (Filardi & Prato, 2018). *Dematerialisation* refers to the devolution of the physical substance of food and the attendant increase of the market value of its immaterial dimensions such as transport, advertising, and distribution. *Digitalisation* includes the morphing of seed and other plant genetic material into patented genetic data, along with the automated robotic technologies transforming rural landscapes, not to mention the incursion of e-commerce into the marketplace where it displaces retailers and restaurants, as in the case of Amazon in India (Guttal, 2018). *Financialisation* refers to the growth of agricultural commodity futures markets driving volatility in food prices. The negative impact of the latter process can be witnessed in land grabbing by foreign investors in countries such as Brazil (Seufert et al., 2018).

The expansion of corporations across credit, inputs, processing, distribution and retail drives agricultural production systems to conform to the needs of global finance capital. This detriments farmers who, as the International Panel of Experts on Sustainable Food Systems (IPES, 2017) reports, are increasingly reliant on a handful of suppliers and buyers, 'further squeezing their incomes and eroding their ability to choose what to grow, how to grow it, and for whom'. Further, dominant firms have become 'too big to feed humanity sustainably, too big to operate on equitable terms with other food system actors and too big to drive the types of innovation we need' (ibid., p. 5). In operating as a means of accumulation for transnational agribusiness companies, financial markets are actively reducing competition within the agricultural industry (Clapp, 2014). In 2013, 175.2 million hectares of biotech products with a market value of \$15.6 billion were grown in 27 countries (Lamphere & East, 2017, p. 75). Nearly half of those products were grown in the United States, where the Monsanto Company uses the poverty of the Global South as a moral platform to justify the public acceptance of technologies. In the last quarter of

2007, while hunger riots broke out in countries including Bangladesh, Cameroon, Haiti and Mexico, Monsanto's profits increased by 45 per cent (Holt-Giménez, 2009).

Companies including Bayer-Monsanto claim that large-scale, conventional agriculture generates higher yields and is more efficient than the small-scale, family farms that produce over three-quarters of the world's food (ETC, 2017). But evidence of the lack of sustainability and resilience of modern industrial farming practices in the form of environmental degradation, high emissions and the double burden of nutrition where obesity meets hunger has led to critical questions about the way we produce food. In a typology that complements food regime theory, three major trends in food systems can be differentiated according to which stakeholders hold power (Holt-Giménez & Shattuck, 2011). The dominant, *corporatist* trend is market-based agriculture driven by transnational agri-food monopolies leveraging policymakers including governments, the WTO and the International Monetary Fund (IMF). The *progressive* trend is based on alternative food networks (AFNs) which work towards environmentally sustainable and socially just alternatives to the corporatist trend including organic and direct-selling relationships between farmers and consumers. The *radical* trend, exemplified by food sovereignty, demands systemic changes to the food regime on the basis of rights, entitlements and redistribution.

These categories have been further augmented to reflect an understanding of food policy in relation to health (Lang & Heasman, 2004). The *productionist* model shares with the corporate paradigm a commitment to raising output, intensifying farming, mass processing, mass marketing, homogeneity of product, monocultures and a reliance on chemical and pharmaceutical solutions. The *life sciences integrated* paradigm claims to deliver environmental health benefits through genetics, biological engineering and laboratory-led solutions while also subscribing to capital-intensive and industrial-scale production to satisfy global demand. In contrast to these top-down, expert-led approaches, the *ecologically integrated* paradigm focuses on citizens rather than consumers. In this respect it is closely aligned with the radical food sovereignty movement. It supports agroecological farming methods based on knowledge rather than external inputs, reduces waste and risk and empowers producers, thus claiming to 'improve links between the land and consumption' (ibid., p. 32).

Each paradigm offers competing, although at times overlapping, views on science, business and consumption. The corporate food regime (McMichael, 2009) reflects the productionist paradigm. Based on the principles of economic liberalism this dominant, hegemonic and market-based model is designed by the G8 governments, global governing institutions such as the WTO, World Bank and the IMF, development agencies and think tanks. Its rules are inscribed in free trade agreements, the US Farm Bill and the EU's Common Agricultural Policy (CAP). Reformist elements can be identified within the corporate food regime. As will be discussed later in Chapter 3, these elements aim for a modification of the current system through the provision of social safety nets, fair trade schemes and organic products that cater to niche markets (Holt-Giménez & Shattuck, 2011). Nonetheless, these solutions are largely technology-driven and rooted in capitalist models of agriculture managed by the same institutions that follow productionist principles and mirror elements of the life sciences' integrated paradigm. The successful rebranding of agribusiness companies such as the Monsanto Company (now Bayer), formerly a chemical producer, as 'life sciences companies' reflects this disturbing mission creep. Both models share a foundation in big business and have the support of government and corporate 'self' regulators. These corporations have huge impacts on the entire food systems of communities where they do business.

A view of the dystopian future

The culture of neo-liberalism imposed on Latin America has made many countries on this continent 'learning labs' or sites of experimentation in the suppression of, and struggle to reclaim, voice. The United Fruit Company, for example, created a built-to-order railway system in Guatemala 'which was ideal if you were a banana trying to get to a shipping port, but rather irrelevant if you were a person trying to get from one place to another in your own country' (Gálvez, 2018, p. 97) In *Eating NAFTA: Trade, Food Policies, and the Destruction of Mexico* (2018), Alyshia Gálvez presents the full horror of free trade unleashed on subsistence agriculture. Twenty years after the initiation of the North American Free Trade Agreement (NAFTA) the livelihoods of thousands of small-scale farmers have been decimated along with traditional ways of eating, particularly the staple corn. Over 300 indigenous varieties have been erased or adulterated by genetically modified varieties, while processed flour produced by heavily-subsidised US farmers floods across the border and economic policies promote concentration of ownership (Mann, 2014a). In 2007, 70 per cent of tortillas and cornmeal consumed in Mexico was produced by Gruma, owned by *Forbes* billionaire 'Tortilla King' Roberto González Barrera. In the same year the price of tortillas rose 50 per cent despite record production levels (De Ita, 2008, cited in Mann, 2014a, p. 100). Once largely self-sufficient, today Mexico imports 42 per cent of its food and possesses poverty rates over 50 per cent. Gálvez characterises the situation as a 'dystopian science fiction':

> … a post-migration Mexico where social engineering has resulted in a low birth rate and the death of the family farm while the poor receive cash incentives for conforming to certain kinds of middle-class behaviour that distance them further from rural ways of life. Local green markets have been replaced with supermarkets overflowing with

> processed food and soda. 'Gringos' clamour for handmade tortillas, while Mexicans have become the world's top consumers of instant noodles.
>
> (Gálvez, 2018, p. 10)

This final point reveals another case of the capture of value – the co-optation of traditional food narratives by global elites. In her analysis Gálvez notes how trade policies have 'taken ancestral ways of eating out of the reach of the average Mexican citizen, while making traditional foods available as a high-value, high-status commodity to be "elevated" and reinterpreted by global elite chefs' in restaurants such as *Cosme* in New York and *Noma* in Copenhagen (ibid., p. 4). The appropriation of capital thus extends to the theft of food stories.

Narrative capital, like all capital, is unevenly distributed. The foundational myths of MesoAmerican people regarding maize, a highly symbolic and sacred food, are effectively stolen and fetishised in high-end restaurants around the world:

> The magical leaf atop a tortilla with mashed black beans that sends [chefs] Redzepi and Bowien into a reverie is avocado leaf, an example of the kind of core ingredient, along with wild greens like *verdolaga*, used by humble rural cooks for millennia in Mexico.
>
> (Ibid., p. 29)

While commercially grown corn is fully commoditised, 'transforming markets, industries, and even bodies as it fuels the production of cheap food and the prosperity of multinational corporations' (ibid., p. 45), landrace corn rises in value. Tortilla consumption in Mexico now declines yearly, dropping 15 per cent in the early 2000s. This is attributed to the increased availability of fast food and lifestyle changes as well as 'status issues' (ibid., p. 47) as urban consumers shun traditional diets for more cosmopolitan foodways to the detriment of their health. McDonald's created an outcry in 2015 when it proclaimed 'Tamales are a thing of the past' but even the strident opposition to this does not mask the decline in consumption of labour-intensive, traditional foods despite their health benefits (ibid., p. 180). The 'holy trinity' of maize, beans and squash is now replaced by what Gálvez (ibid., p. 148) calls the 'salmon and broccoli approach' which promotes specific diets over culturally appropriate, traditional fare. She notes that native 'superfoods' such as amaranth and chia are not promoted by government campaigns while they appear on the shelves of supermarkets throughout the West.

Meanwhile the 'modernisation' of the Mexican economy including the liberalisation of markets has created an import-dependent nation with lifestyle-related illnesses of epidemic proportions. The country's multi-sectoral National Strategy for the Prevention and Control of Overweight,

Obesity and Diabetes – the latter being the number one killer in Mexico – includes a soda tax (alongside regulations on food marketing, an anti-poverty programme called Prospera and the campaign 'Crusade against Hunger'). Public health responses frame citizens as consumers responsible for their own choices; therefore diseases such as diabetes could not be the fault of political forces but instead are due to the lack of education and poor decisions of eaters themselves. The market is again conjured to solve not only poverty and inequality but disease. Despite growing awareness and mounting evidence of the impact of calorific, discretionary foods on Mexican health, Big Food is not reducing production of these foods. Instead corporations are setting their sights on markets in the Global South such as Mexico and using their considerable marketing resources to promote their products. Spending on advertising far exceeds the cost of ingredients and processing cheap calories. In comparison to cosmopolitan fast foods traditional foods are perceived as unmodern and embarrassing, and the 'children cry for Burger King' (Schortman, 2010). Given the lack of 'narratives of hope, semiautonomous cultural spheres, discussions of meaningful differences, and non-market-based democratic identities' for children in the media (Giroux cited in Couldry, 2010, p. 126), is it any wonder that these attitudes emerge?

The preservation of Mexican cuisine is, ironically, a global priority; it was inducted into UNESCO's List of Items of Intangible Heritage in 2010. Regrettably, this recognition does not come with support for those who grow and produce the cuisine to sustain their livelihoods. The romanticisation of rituals, culinary techniques and ancestral customs are accordingly consigned to a museum, objects of curiosity. Much like the indigenous custodians of these practices whose voices are elided in the retelling. It is in these omissions that the value of food is misconstrued. Roberto González (cited in Gálvez, 2018, p. 37) refers to the 'gamut of activities related to food production and consumption' that do not distinguish between Mexican agriculture and eating but represent an effort to sustain households and families across generations. *Milpa* cultivation, for example, a form of agroecology based on small-scale production of cycles of corn, squash and beans, is more than a way of improving nature; rather it is 'a way of getting closer to the processes of nature getting as close as possible to what nature does' (Jésus León Santos cited in McMichael, 2013, p. 139). It works on the basis of 'ecological capital' as an alternative form of value where co-production with nature including engagement in pluriactivity increases farmer autonomy by mitigating risk, lowering costs, and ideally, avoiding debt to agro-input suppliers. Ecological capital, McMichael argues, is analogous to food sovereignty, invoking 'value-adding' as the goal and outcome of peasant agriculture as conceptualised in van der Ploeg's (2008) theory of re-peasantisation. This clearly demarks the difference between a food system and a cuisine.

The fetishisation of Mexican food can be interpreted as another example of how 'the powerful have at each stage taken time and energy to record and remember that which they are busy trying to eliminate' (Gálvez, 2018, p. 36). It may seem extreme to compare the historicising of indigenous people through *indigenismo*, as exemplified in Diego Rivera's iconic murals, with the hand-grinding of landrace Mexican corn by world-class chefs for well-heeled diners in Manhattan, but it is no coincidence that traditional Mexican foods are now in vogue when two decades of integration into the neo-liberal trade regime has ruined small farmer livelihoods and displaced millions. A different dynamic, with a similar outcome, operates in the decline of local consumption of quinoa, an Andean staple now fashionable as a source of healthy protein in the cities of the Global North. Today, many Bolivian and Peruvian farmers opt to sell their entire crop and eat white rice imported from China to meet the preferences of elite foodies in the West. Whether or not the heritage corn of Mexico is heading for what Gálvez calls a 'quinoa moment' it is too late for many grain growers in the Mexican countryside.

With startling hypocrisy, elite dining steps out of the conventional food system by re-inserting values of authenticity in the form of signifiers such as geographic specificity, simplicity, personal connection, history, tradition and ethnic connection (Heldke, 2005 cited in Johnston & Baumann, 2010, p. 73). This reinforces the colonial notion that 'the Other as a cultural resource will always be available to be consumed by more powerful members of the geo-political core' (ibid., p. 102). Food justice and food sovereignty are discursive omissions in the gourmet foodscape when, in fact:

> If place matters, in theory, so do the particular political histories of control over that place, and oftentimes, struggles for rights, legitimacy, resources, and domain.... *Terroir* is just one more in a large toolkit of words for distinctions that are value added to products, enabling some products to stand out from a sea of others that are, in other senses – such as nutritional or use value – identical and interchangeable. When a food is elevated to a *cuisine*, it is not about its power to fuel the human body, or even its capacity to bring people together around a table, but its function as a vehicle of value that enable some to profit wildly by it.
>
> (Gálvez, 2018, p. 44)

This reveals a contradiction in our 'democratic openness to discover new food cultures, alongside the continued existence of exclusion, inequality, and exploitation in the gourmet foodscape' (Johnston & Baumann, 2010, p. 2). Josée Johnston and Shyon Baumann (ibid., p. 3) describe foodscapes as 'dynamic social constructions' that relate food to particular places, people and meaning that can thereby 'capture or obscure ecological origins and social implications' of production and consumption (ibid.). They claim that in the 'omnivorous era' two frames dominate – authenticity and exoticism. The ability to buy and sell authenticity, 'generated through perceptions of how a cultural object negotiates a set of standards and values, instead of emerging from a cultural object's qualities' (ibid., p. 71), demonstrates how market culture and consumerism facilitate the capture of value and provide the basis of democratic ideology under neo-liberalism.

Feeding the world

Narratives of food as a commons, including food sovereignty and practices of agroecology, are about equal and proper valuing of the multiple dimensions of food, while the production of cheap food has always required the 'suppression of political dissent' (Patel & Moore, 2017, p. 151). These methods of suppression – some covert, others less so – include the corporate capture of nature and knowledge practices legitimised discursively by multinational agribusiness and life science companies purporting to feed the world. In an ever-shrinking corporate landscape Bayer & Monsanto – 'Two companies ... Now, Advancing Together as one' (Bayer AG & The Monsanto Company, 2018) – employ visions of an onrushing dystopian future and crippling poverty as a moral platform to establish the necessity of biotechnological intervention. (Bayer has dropped the Monsanto name for reasons that will become clear in the potted history of Monsanto Company to come later in this chapter). This US$88 billion 'marriage made in hell' (AFP, 2016) is the latest in a series of mega-mergers including DowDuPont – now Corteva – and ChemChina-Syngenta, which put two-thirds of the commercial seed and agrochemical markets into corporate hands – and coffers (IPES, 2017). Monsanto has been particularly aggressive in countering criticisms against the introduction of its technologies, including genetically modified organisms (GMOs), by employing strategies such as 'be part of the conversation' campaigns on social media encouraging the public to interrogate the company's practices (Lewis, 2018). Bayer also pacifies customers with mediated transparency through interactive web portals that claim to provide dialogue with experts. In 2017 it launched the Crop Science Transparency Initiative to provide access to crop protection studies via a dedicated platform. The platform was awarded Best Public Outreach Program at the industry's Agrow Awards (Bayer, 2018a). The following quote, from Bayer's helpful video 'Understanding regulatory science for crop protection (#lovethefacts)' is designed to help us negotiate the growing complexity of our foodscapes:

> People worry about pesticides in their food, their health, the health of bees and the environment. Conflicting messages dominate media and social networks, making it difficult for people to separate fact from fiction. Mistrust is amplified when the public doubts the integrity of

> privately-funded research studies, or if they think the industry is hiding something from them.... This one goes out to the fact guys. To the curious, the decisive ones, to the ones eager to learn ... every little detail and every little fact. But when it comes to some of the most important issues on our planet, this is often getting tough and complex and confusing. Because you don't know who to trust and where to find the real proven facts. That's why we want to talk, talk about crop protection, pesticides and plant medication. About the environment and our health and growing food for a growing world population.
>
> (Bayer, 2018b)

In this presentation Bayer's Fact Guy is informed by a stable of lawyers, safety experts, regulatory managers and documentation specialists to whom you can direct questions, online ('please remember, there are more of you than us, so it may take a while for us to answer your questions'). Previous concealment of information is justified as 'the protection of our intellectual property, including our own safety research'. Attempting to dispel the idea that the decision to selectively reveal safety reports is profit-driven, Bayer explains this uncharacteristic openness is 'not to sell you a product, but to connect with you and share our beliefs'. The site provides complete studies on controversial product ingredients, such as Flupyradifurone, Mesosulfuron, and Propineb, as well as articles on the safety of pesticides in the environment and our food. While reports on the various products are presented in their original, prolix form without any translation of highly technical language, articles on topics such as 'Is our food safe' are presented in simple text and comical videos.

Bayer (2018b) justifies selling pesticides for crop growth on the grounds that it's 'not because we're evil, but because deep in our curious engineer hearts we know that leads to better solutions to provide enough food for everyone'. Acknowledging that there are 'residues of pesticides in our food' Bayer presents 'real proven facts' that these aren't toxic. In order to justify our ability to safely consume foods with pesticides, Bayer argues that 'our body handles a whole bunch of chemicals every day', likening pesticide residue to coffee in that one would have to consume an extreme amount of each in order for the toxins to have a noticeable effect on the body. 'Everyday' items we consume that are toxic or carcinogenic include 100 cups of coffee. Good to know.

Framing scarcity

The justifications presented by Bayer and Monsanto exemplify the battle between competing frames, stakes and forms of expertise that is escalating with concerns about the resilience of global and local food supplies. Trading on notions of food scarcity in an uncertain future these companies and their competitors – a shrinking pool – have successfully embedded a positive view of biotechnology in the public mind for decades. This embedding is enabled by what van Dijk (1998) refers to as 'legitimation strategies': discursive resources that are employed to frame issues and technologies in a deliberate manner and convince audiences to adopt these intended representations. Legitimation is an ongoing process evident in corporations' everyday practices. In the arena of food and agriculture these practices construct an ideological space within which the issues and developments surrounding biotechnology can be publicly received and supported. Consider, for example, the framing of the Bayer-Monsanto merger by Liam Condon, Member of the Bayer Board of Management of Bayer AG and President of Crop Science that 'today's milestone means that the two leading innovators in agriculture will now come together as one to shape agriculture through breakthrough innovation for the benefit of farmers, consumers and our planet' (Bayer AG & The Monsanto Company, 2018). The altruistic aim of 'shaping agriculture' for the common good of humanity is integral to discourses that justify transgenic technology. At the core of justification for biotech products is the idea that they are critical for guarding humanity against future disaster in the form of climate change, the scarcity of natural resources, overpopulation and growing poverty in the Global South.

Paradoxically, obesity has eclipsed hunger as the most widespread global health problem (NCD Risk Factor Collaboration, 2016). It is in its dialectical pairing with abundance that the relational and socially-constructed nature of scarcity emerges. Framing scarcity as an inherent characteristic of productive resources such as soil, water and seeds naturalises it, and effectively sidesteps issues of social justice that run counter to the interests of elites (Mann, 2017a). Food scarcity is not 'natural' in the sense it is naturalised in the discourse of the market. It is a 'concrete period of dearth' (Mehta, 2011, p. 382) felt acutely by a rural population when a productive resource, such as water, is limited. In response farmers adapt. On a daily basis they apply strategies rooted in local knowledge systems and practices that deal with seasonality and uncertainty in weather conditions. These highly differential coping mechanisms are designed to respond to the lived experience of scarcity, or coping with the 'regularly irregular', such as rainfall (ibid., p. 379).

Advances in biotechnology are constantly cited as improving productivity and quality in both livestock and crop yields despite claims that 'business as usual is no longer an option' for ensuring food security (IPES, 2017; Copeland, 2016). As political ecologists Richard Peet, Paul Robbins and Michael Watts (2011, p. 38) remind us science is itself 'a highly problematic global political enterprise' which is 'never conducted entirely separately from the global political and economic forces that make it possible'. Further, the domination of science as arbiter of environmental 'truth' has led to the marginalisation and overshadowing of 'different ways of knowing and explaining the world', thereby 'putting power in the hands of technical experts' (ibid.). Neo-liberal discourses frequently 'underplay' rational concerns regarding the safety of technologies to human health and the environment, claiming that 'there is no evidence of harm from industrial technologies or that newer, more efficient technologies will replace the old ones soon enough' (Holt-Giménez, 2017, p. 176). In relation to discourse on climate change Massimo De Angelis (2017, p. 1) notes

> silence and illusions are packaged as progress; there's no cause for alarm after all, we are often told, we will one day hire engineers who will design big vacuum cleaners to suck up all the excess carbon dioxide and store it in old oilfields and build big walls to protect the richest cities.

Boaventura de Sousa Santos claims that scientific knowledge is a form of 'globalised localism' and our current ecological crisis is based on the fact this knowledge is situated in and 'perpetuates the relations of colonial inequity, giving shape to a monoculture of knowledge' (Santos, 2007, p. xxxix). The rise of 'alternative knowledges' challenges this conception by demonstrating the hybrid and situated nature of knowledge, as the outcome of 'socially organised practices involving the mobilisation of different types of material and intellectual resources bound to specific situations and contexts' (ibid.).

The 'non-political' status of science is contested by concerns over the increasing power of corporations at the expense of small-scale farmers even in developed and highly urbanised countries (Australian Institute of Agricultural Science and Technology, 2016; Mann, 2017b). The continued expansion of transnationals was heavily critiqued by Hilal Elver, the UN special rapporteur for the right to food, in 2018. In her report on the practices of global corporations manufacturing pesticides she states that companies are guilty of 'systematic denial of harms' and 'aggressive, unethical marketing tactics' (Elver, 2018). Condemning lobbying practices that have 'obstructed reforms and paralysed pesticide restrictions', Elver accuses corporates of infiltrating federal regulatory agencies via 'revolving doors' and cultivating 'strategic public-private partnerships that call into question their culpability or help bolster the companies' credibility'. She goes on to argue that this credibility is bolstered by academics and regulators who accept corporate funding and sign confidentiality agreements, sacrificing autonomy and engaging in disinformation campaigns that support Big Ag agendas. Room for debate over the science presented by companies is limited in these conditions and therefore, as Darrin Nordahl (2014, p. 178) notes 'unless food consumers are part of the family corporations of industrial agriculture, or sit on the boards of the larger, public corporations, they have little voice in food choice'.

Shifting discourses

It is important to consider a company's business history in order to identify the strategy behind any shifts in discourse (Glover, 2010). Monsanto Company, for example, was originally a chemical company primarily producing polychlorinated biphenyls (PCBs) used in electronics and machinery as a coolant, along with several other products for the food and pharmaceutical industries. For decades Monsanto products were plagued with questions over safety. These included the artificial sweetener saccharine which the US Department of Agriculture attempted to ban due to concerns it was a carcinogen; polystyrene which was restricted in several US cities not only because it isn't biodegradable but also because of evidence that styrene – the key ingredient – was carcinogenic; and dioxin and Agent Orange, which later were used in military warfare and banned in several countries. In 1979, PCBs were banned in the US after their toxicity was acknowledged by government (Lamphere & East, 2017). Monsanto knew as early as 1937 that PCBs were highly toxic, and yet carried on production. This resulted in a successful legal action on behalf of the 24,000 people living in Anniston, who argued that Monsanto's nearby manufacturing plant knowingly contaminated the town with the carcinogen (Robin, 2008, pp. 9–11). Along with the pressure of court actions and product bans, Monsanto's move towards biotechnology or 'life sciences' was shaped by a number of developments that had 'no direct connection with sustainability or poverty alleviation in the developing world' (Glover, 2010, p. 71). These included the rising cost of raw materials in the petrochemicals business and the growing environmental movement threatening smear campaigns and legal action against the petrochemical industry.

A political opportunity for the commercial exploitation of biotechnology and genetic modification was provided by the 1973 genetic engineering breakthrough by Stanley Cohen and Herbert Boyer followed by the US Supreme Court ruling that transgenic life forms could be patented. By the mid-1990s, biotechnology was declared Monsanto's strategic focus and investments expanded into the seed industry (Glover, 2010, p. 73). The company released its most successful product to date, Roundup Ready, which Monsanto argues was 'a perfect fit with the vision of sustainable agriculture and environmental protection' (Monsanto Company, 2014a, cited in Lamphere & East, 2017, p. 80). After the patent on Roundup Ready expired, Monsanto introduced Roundup Ready crops to the market, allowing it to maintain its dominant share of the glyphosate market (Schluter, 2015, p. 6). The crops were genetically modified to withstand the effects of Roundup Ready pesticide, thereby giving farmers the ability to spray fields against weeds without damaging the crops, and giving Monsanto control over the emerging biotechnology market (ibid.). In 2017 the chemical was relicensed in Europe despite a petition signed by 1.3 million people and the World Health Organisation's 2015 announcement that it was 'probably carcinogenic to humans' (Nelsen, 2017). In a recent ruling the company was ordered to pay school groundskeeper Dwayne Johnson US$289 million for failing to correctly label the product and therefore being responsible for his cancer (ABC News, 2018). Bayer shares have declined in value more than 10 per cent since this ruling (Reuters, 2018).

Post biotechnology: shifting purposes

Monsanto was tasked with restructuring its scandal-ridden chemical corporation into a life sciences company in the mid-1990s. In order to create this distinction in company operations Monsanto needed to strongly promote its commitment to values like globalisation, sustainability and culture. An apocalyptic narrative emerged, one that was 'most efficient at generating and legitimating massive society-wide sacrifice', and as such 'the only narrative form that can sustain war as culturally acceptable' (Smith, 2005, p. 27). While Monsanto was not heading to war, it was set to battle both its image as an unethical chemical production line and community concerns over emerging life sciences technology. By invoking images of a starving world and painting a Malthusian picture of overpopulation, Monsanto could convince a sceptical audience that heroic intervention, in the form of '... a different model of development ... and brand-new technologies' (Monsanto Company, 1995a), was necessary to ensure the future of humanity.

Monsanto also needed to convince its stakeholders and employees of the corporate shift towards biotechnology. To do so its leaders employed a lexical field that reflected neo-liberalism's political and economic primacy in the Global South and Global North. Machin and Mayr (2012) define a lexical field as word choices signifying a particular discourse that reflects certain norms and values. In corporate communications, Monsanto intercut discussions surrounding global sustainability with phrases like 'growth of new areas' (Monsanto Company, 1995a), 'global marketplace' and 'cultural innovation' (Monsanto Company 1995b, 1996). Employing neoliberal discourse that favoured commodification, open competition and free-market globalisation allowed Monsanto to link corporate strategy with sustainability, positioning its corporate stakeholders to see the life science direction as legitimate.

Not all senior members of Monsanto espoused the company's line about biotechnology's environmentally sound applications. In a 1992 paper, the company's former Executive Vice President and Chief Technology Officer Robb Fraley said genetic modification had the potential for providing 'readily accessible, economically viable means of addressing primary food production needs' to feed the world's growing population (Fraley, 1992, p. 40). However, he conceded genetic modification was not a quick fix for global hunger and cited a paper which cast doubt on whether 'tinkering' with technologies designed in and for the 'developed' world could help the 'developing' world (Hodgson, 1992). Glover argues

> the fact that Fraley apparently acknowledged those points is a fascinating indication of the cognitive dissonance opening up between the kinds of technologies Monsanto was actually developing and the company's claims about their relevance to the needs of farmers in the developing world.
>
> (Glover, 2010, p. 77)

In the early 2000s Monsanto moved on from fostering a positive image as a biotechnology company to fostering a new image for biotechnology itself. Global concerns around the safety and effectiveness of biotechnology were rising, as indicated by governmental responses to a campaign from Monsanto at the end of the twentieth century that envisioned biotechnology as 'feeding the world'. Delegates to the FAO on Plant Genetic Resources released a statement inviting European and African nations to stand in solidarity against images of their poor and hungry citizens being used 'by giant multinational corporations to push a technology that is neither safe, environment friendly, nor economically beneficial to us' (Delegates to the FAO on Plant Genetic Resources, 1998). European food retailers started pulling foods containing GM ingredients from the shelves, and within months the European Union imposed a de facto moratorium on further approvals. An analysis from the Deutsche Bank concluded that 'GMOs are Dead' (Brown & Vidal, 1999).

Up until this point, Monsanto had relied on negative representations of nature as a justification for biotechnological intervention (Monsanto Magazine, 1962). Afterwards, however, Monsanto started portraying biotechnology as simply building upon 'the natural advancement in the progress of agricultural science' in order to placate public concerns that it was harmful (Lamphere & East, 2017, p. 82). At the time, a web page outlining Monsanto's pledge included a link to a transcript of President Hendrick Verfaille's speech:

> To put agricultural biotechnology into context: humans have been planting and improving crops for 10,000 years. Today, almost everything we eat is the result of improved animal and plant genetics over time. In the last 200 years – from Gregor Mendel through Watson and Crick's discovery of DNA to today – we are learning how biology and genes work. We can now be very precise and efficient in creating new varieties and traits, and in how we farm.
>
> (Monsanto Company, 1999, cited in Lamphere & East, 2017, p. 82)

The low-mimetic narrative embedded in this mundane description of business as usual normalised the technology while Monsanto also worked to build its image as a company concerned with global sustainability. The company engaged in a concerted Corporate Social Responsibility (CSR) strategy, producing the 'New Monsanto Pledge' which featured five major commitments: dialogue, transparency, respect, sharing and benefits 'commercially as well as environmentally'. Farmers in the Global South were directly implicated in Monsanto's 'sharing' commitment to help improve food security and protect the environment by 'bring[ing] the knowledge and advantages of all forms of agriculture to resource poor farmers in the developing world' (cited in Glover, 2007, p. 21). As Machin and Mayr (2012) argue the word 'commitment' suggests a mental assurance but conceals the performance of any actual action. Nevertheless, this discourse was designed to create an association between Monsanto products and environmental sustainability in the minds of audiences. The Company's declarations of CSR were arguably less a reflection of concerns for environmental sustainability and more about ensuring its business' longevity.

Since 2005 Monsanto has been tasked with neutralising growing public outcry against biotechnology and GM foods. It has tried to strengthen links with sustainability by funding research and creating public outreach programmes. One of these projects is geared towards protecting the honey bee, 'vital to the production of fruits, vegetables and nuts we enjoy every day' (Monsanto Company, 2016). Critics have argued in recent years that Monsanto's herbicide is responsible for killing bees and destroying the natural ecosystem (Vallianatos, 2017). By investing in research to protect the honey bee (Monsanto Company, 2016), Monsanto not only combats these concerns but more broadly frames the company, and all its work, as altruistic. Monsanto cements this perception by incorporating a human rights section in its corporate pledge to 'create a metric to which we will hold ourselves accountable and [that] demonstrates our commitment to human rights in the conduct of our business globally' (Monsanto Company, 2006).

The onrushing future

For the past four decades, Monsanto has relied on the 'onrushing future' as a justification for their products. This metaphor communicates an urgency, whereby

> we are not moving to the future, but the future approaches us ... the future is in a hurry. There is no time for fundamental questions. If we wait too long before acting according to the prophecies, the future will override us with new disasters.
>
> (Jansen & Gupta, 2009, p. 437)

Imageries of the future provide a discursive basis for examining the costs and benefits of technology. The 'beckoning' future envisions a positive future (de Wilde, 2000, cited in Jansen & Gupta, ibid.). It doesn't negate the possibility of things going wrong, but emphasises the role of human agency in choosing the right path. Typically, the beckoning future can only be realised if society behaves according to conditions prescribed (Jansen & Gupta, ibid., p. 473). By contrast, the onrushing future leaves us with little choice. It is invoked by Monsanto to spur consumers into rapid, unquestioning action.

Monsanto first employed imageries of an apocalyptic future to counter criticism in the 1960s, releasing *The Desolate Year* (Monsanto Magazine, 1962) in response to scientist Rachel Carson's *Silent Spring* (1962), which linked the products of chemical companies like Monsanto to environmental toxicity and other negative effects. In no ambiguous terms, the Company pamphlet set out to answer one question: 'Does the world really need chemical pesticides?' (Monsanto Magazine, 1962). Using a narrative structure and highly descriptive turns of phrase, *The Desolate Year* painted a dystopian future where crop-threatening insects have been allowed to run wild without pesticides, destroying humanity's food source. Monsanto warned:

> The bugs were everywhere. Unseen. Unheard. Unbelievably universal. On or under every square foot of land, every square yard, every acre, and county and state and region in the entire sweep of the United States.... So went the fresh, clean vegetables. So went sweet corn.... If its stalk and ear escaped the harsh attack of the borers, along came the earworm, hatching from eggs that a brown-grey moth slipped into the receptive silks alongside the life-giving pollen.
>
> (Monsanto Magazine, 1962, pp. 4–5)

The pamphlet's conclusion warns the horrific series of events leading to the world's destruction in the story have all previously occurred in real life, albeit locally and to a lesser degree. Highlighting this fact satisfies a dual purpose; it drives home the *possibility* of the beckoning future, should we act appropriately, and promotes discontinuity with the past (Jansen & Gupta, 2009, p. 37). Monsanto continued to frame nature as an untamed threat throughout the late twentieth century. In response to challenges about their environmental pollution record in the 1970s, they released the slogan: 'Without chemicals, life itself would be impossible' (Monsanto Company, 1979).

While Monsanto has continued to invoke the onrushing future metaphor to justify its products, the threat to food security portrayed has shifted from insects and the natural environment to play off evolving concerns over climate change and global population growth. In a speech delivered in 1999 at the State of the World Forum, Monsanto's CEO addressed the impending environmental crisis, stating:

> Five point eight billion people in the world today, and one and a half billion of them in abject poverty. About one person in seven, about

> 800 million people in the world, so malnourished that they cannot participate in work life or in family life, living on the edge of starvation.... In the next 30 to 50 years, we will have a world population of somewhere between eight and twelve billion people. There is no more land. The land has been used.
>
> (Monsanto Company, cited in Lamphere & East, 2017)

Monsanto began to employ the socially-constructed notion of food scarcity (Mann, 2017a) throughout the 1990s, particularly in its 'Let the Harvest Begin' campaign in Europe. This narrative, circulated through major European newspapers, threatened a world of famine due to a massive increase in 'third world' populations. It represented biotechnology as the only solution to this onrushing ominous fate, arguing that it will 'play a major role in realising the hope we all share' and 'accepting this science can make a dramatic difference in millions of lives' (Bruno & Greer, 1996). Monsanto continues to implicitly blame growing and impoverished populations for future food shortages. As Schluter (2015, p. 31) observes, Monsanto's website, social media accounts and various PR campaigns demonstrate 'the view of the world from Monsanto is a grim one, where people are already starving and more will starve without Monsanto's intervention'.

At first glance, Monsanto's language could suggest it has shifted to enticing consumer support with visions of a beckoning future. Monsanto's website greets us with the statement: 'In the face of a changing climate and other environmental challenges, we're helping ensure our agricultural system continues to suit the needs of everyone' (Monsanto Company, 2017). This suggests a positive outlook as long as we follow Monsanto's prophecy or, rather, use their products according to packaging guidelines. Similarly, in line with the beckoning future metaphor these products are implied to be giving us back autonomy over the future. For example, in the introduction to its 2017 annual Sustainability Report, CEO Hugh Grant described food insecurity as 'a wide range of challenges' and framed the company's goal in terms of 'helping create solutions to these challenges while helping to take care of our planet' (Grant, 2017). The plurality of options suggests choice and the ability to exercise it, luxuries not afforded in the wake of an onrushing future. It is implied that 'a direct relationship exists between technology and the reduction of poverty, that simple availability of new technology will improve the situation of the rural poor' (Jansen & Gupta, 2009, p. 440). Indeed, what remains absent from Monsanto's imageries of the future is any explanation as to how their products will end hunger in food insecure nations. As Schluter (2015, p. 31) says, 'it almost appears to be unimportant to Monsanto how their products fit into saving the world'.

Monsanto also frames biotechnology as the answer to consumers' racialised fears that 'overpopulated' countries will consume the world's food supply. Over the last few decades, 'the bodies of third world populations take the place of those rapacious insects. People of colour emerge as both the source and the "victim" of food shortage, which is ascribed to natural, not political cause' (Patel et al., 2005, p. 432; see also Mann, 2017a). By exploiting racialised collective fears of ecological apocalypse, this discourse conceals gaps in reasoning and justifies an argument that doesn't follow from the premise. Further, Monsanto fails to acknowledge that biotechnology is a double-edged sword for small-scale farmers. Farmers must 'get big or get out', urging family farmers to adopt biotechnology and use expensive inputs or leave the land. In many cases Monsanto's contracts with farmers also prohibit them from saving seeds year-to-year, requiring them to purchase new seeds each time (Kucinich, 2017). These seeds are genetically modified with gene-use restriction technologies to produce a toxin that renders the seed sterile – able to produce a plant and be harvested for food but unable to reproduce. This technology was rebranded by activists concerned about the commercialisation of a product that would make farmers dependent on biotechnology companies as 'terminator'. In 1999 Monsanto's chief executive promised not commercialise terminator seeds. This promise was not kept and the Company continues to sue farmers, such as 75-year-old Indiana soybean farmer Vernon Bowman, who violate their patents by re-sowing seeds season-to-season (Ledford, 2013). Paradoxically terminator technology has begun to appeal to environmental advocates concerned about the contamination of organic farms by genetically engineered crops, particularly those engineered to produce chemicals and pharmaceuticals.

By employing the language of winning and losing to frame a world with and without biotechnology Monsanto attempts to suppress opposition and thus 'neutralise' the market (Jansen & Gupta, 2009, p. 437). Questions of how agricultural markets are socially shaped and who has access to technologies are omitted. Yet Monsanto's discourse is persuasive because it 'requires very little critical analysis from its subscribers' (Schluter, 2015, p. 34). This appeals to 'cognitive misers' (Weasel, 2008) who welcome cognitive shortcuts (i.e. seals of approval) to make decisions regarding safety. In short, the cognitive miser is one who does what makes one's life easier. Schluter (2015, p. 34) argues that consumers don't question Monsanto's dominant world view because it 'is easier than questioning every packaged food in the US, and this makes the dominant discourse extremely palatable'.

A new, legitimising mantra is emerging in the Fourth Industrial Revolution. Where advocates of the Green Revolution sought to feed the world, the new order seeks to build 'inclusive and sustainable food systems based on new technologies' (Filardi & Prato, 2018, p. 6). 'Precision agriculture', a vision of uber-mechanised agriculture driven by digital technologies and algorithmic processing of Big Data, is the new frontier of agribusiness (Monsanto Company, 2018a, 2018b). Monsanto moved into this brave new world with the acquisition of digital agriculture subsidiary Climate Corporation in 2013 and, prior to the merger with Bayer, had once again reinvented itself – this time as a 'data robo-tech company' (Zundel & Ribeiro, 2018, p. 28). Meanwhile Bayer firmly entrenched itself in the field with two key purchases – Zoner, a Canadian company specialising in satellite and aerial imagery and data on yield and soil conductivity that provides real-time weather information, in 2015, and; proPlant, a German plant health diagnosis system, in 2016. Climate Corporation was a significant factor in the Bayer-Monsanto merger according to industry reports, and it is busily buying up precision agriculture start-ups including GPS-based information systems for irrigation and soil analysis. The global farm machinery sector is another area experiencing oligopolistic concentration.

These trends not only reduce farmer autonomy but make possible industrial farming on so-called 'marginal lands' where many small-scale farmers in the Global South live and produce food. This will extend the land grab already in train, and further threaten the livelihoods of millions of farmers and farm workers who will have no role in an automated farming future (Zundel & Ribeiro, 2018). Further, claims that these technologies may reduce chemical inputs through enabling smaller and targeted applications do not address the other elements of sustainable farming that include improving soil health and building climate change resistance through agroecological multispecies planting.

Who speaks for nature?

Campaigns against the genetic erosion that accompanies the corporate capture of biodiversity illustrate contention over not only nature itself but rival forms of knowledge. Biodiversity is defined by the United Nations Convention on Biological Diversity as 'variability among living organisms of all origins, including, inter alia, the terrestrial, marine and other aquatic ecosystems and the ecological complexes they are part of ... including the internal diversity of species, between species and of ecosystems' (cited in Santos, 2007, xli). It also includes knowledges about biodiversity, recognising it as a field comprised of 'a set of discourses where different knowledges, cultures, and political strategies intersect' (ibid.). In the neo-liberal economy multiple institutions including northern NGOs, research institutes, universities, pharma and agribusiness companies seek custodianship of these knowledges, including an array of different cosmologies and languages. These attempts at co-optation are opposed by counter-hegemonic movements in the Global South. They contest the global-centric view of biodiversity as a resource to be managed and financialised in forms of 'market environmentalism' (Peet et al., 2011, p. 24) with 'bio-democracy', a term that refers to the radical redefinition of production away from uniformity or monocultures to diversity, emphasising threats to biodiversity including development, productivist agriculture, and habits of consumption governed by narrow economic models (Escobar, 1998). Given the abundant evidence of the failure of 'capitalism unleashed' to protect forests, seas and the atmosphere, 'the commodification of nature to save it' is 'contradictory at best' (ibid.). Instead these processes have created sources of 'biocapital' – forms of value 'based on the commodification of the building blocks of life itself' (Holt-Giménez, 2017, p. 188). Capitalism is thus revealed to be 'a translation machine for producing capital from all kinds of livelihoods, human and non-human' (Tsing, 2015, p. 133). This machine functions on the basis of 'alienation' as a form of 'disentanglement' through which life forms such as seeds and their genetic material are 'removed from their lifeworlds to serve as counters in the making of further investments'. These processes of alienation make possible accumulation or the amassing of capital – converting 'ownership into power' (ibid.).

LVC critiques biodiversity itself as a hegemonic construction, noting that the discourse of biodiversity can be deployed to limit cultural autonomy and disable livelihoods in specific places. The majority of biodiversity exists in territories inhabited by indigenous peoples for whom nature cannot be divorced from society; their cosmologies do not divide and categorise the world like Western science which codifies biodiversity through instruments such as the agreement on Trade-Related Aspects of Intellectual Property Rights (TRIPS). Just as monopolies of patents can enforce a regime of monocultures 'those with capital can overturn communities and ecologies' (ibid.) in other, diverse ways; not only through privatising and pricing but through *protecting* natural resources. These examples of the capture of common property resources demonstrate how 'capitalist forms can flourish even across great circuits of difference' (ibid.). Indigenous people themselves are now 'genetic resources and tools for accessing, through traditional knowledges, plants and other living beings' (Santos, 2007, xiv). This and other forms of biopiracy are being countered by alternative frameworks for legislation such as the Model Proposal of Law of the Organisation for African Unity (Third World Network, 1998) seeking to promote the protection through law of biodiversity and local forms of society organisation to secure food sovereignty and rightful custodianship of 'rival knowledges' that contest the corporate paradigm.

Conclusion

This chapter has explained how multinational interests irrevocably change traditional foodways and seek to suppress the voices of those most affected by those changes. The production of biocapital not only captures and commodifies value but 'unleashes the power of speculative capital investment' making a 'genetically modified future' highly profitable for a growing network of scientists, technicians, investment companies, development agencies, public relations consultants, and agribusiness firms (Holt-Giménez, 2017, p. 190). These stakes drive companies to take radical action to maintain their dominant discourses. Some of this action has been reactive against critics of the corporate food regime who expose food scarcity as a sociopolitical construct that legitimises further intensification of production at the expense of people and environments. These critics include advocates of food sovereignty which resists simplistic linkages between population growth, climate change, conflict and resource scarcity, and reminds us that technological solutions are not neutral; that the food industry itself is a major contributor to overproduction, food insecurity and environmental degradation; and that unaffordable housing, low incomes, disability, and poor access to transport are among the complex factors that contribute to alarming rates of food insecurity even in comparatively wealthy economies.

Food sovereignty 'pushes back against capitalism' by insisting that food is treated 'not as a commodity but as a common good' (Vivero-Pol, 2017, p. 17). The continuous struggle between these paradigms is, in a sense, inevitable as the commons and capital

> cohabit the social space, their struggles and relative powers giving shape to it, with the result that unevenness and contradictions are many, as well as strategic games to colonise the other's space with one's own values and decolonise one's own space from the other's values.
>
> (De Angelis, 2017, p. 276)

Complexity is found in small groups in communication hence the need for 'frameworks that allow for a continuous flux of argument and a diversity of beliefs and values, kept together by the rhythm of our daily reproduction and the decisions that need to be communally taken in given contexts' (ibid., p. xiii). The creation of these frameworks in the form of inclusive models of participation and governance is the focus of the following chapter.

3 Recovering and reclaiming voice

The crisis of our food system is fundamentally a crisis of democracy. This means that questions of fairness and affordability are more-than-food questions; they are 'questions of justice and equality and rights, of enhancing rather than restricting democracy, of making a more rational, legitimate economy' (Bittman, 2004). There is a need for more democratic governance of food systems that accommodates 'alternative perspectives on food and its value to society' and addresses 'the differential power and political interests associated with different perspectives on food values' (Eakin et al., 2010, p. 263).

So, how does one 'do democracy' in the context of food systems? Tim Lang (2005) argues that putting food at the centre of democratic culture has to be fought for. Being passive spectators or eaters cedes control of food environments to the large, profit-driven multinationals that dominate markets. This aligns with wider critiques of a democratic deficit in countries such as Britain where the neo-liberal consensus (that is, privatisation, abandonment of social welfare, and the prioritising of flexible labour markets at the expense of livelihoods), a lack of effective political actors and the decline of a vibrant civil society 'foreshorten' the democratic process (Couldry, 2010, p. 48). Communicative trust between the political class and citizens is low in a system that 'provides formal voice for its citizens but *fails so markedly to listen*' (ibid., p. 50).

This leads to another crucial question: can food environments be democratised in spite of an imperfect democratic system? Food democracy means 'enhancing the role of citizens in the management and control of the food system' (Renting et al., 2012, p. 293). Participation, ideally, incorporates emancipatory learning to support equitable and inclusive civic engagement. In this chapter I problematise participation, through consumerism and more generally, before considering how principles of a 'commons resilience' can be built into processes of planning and governance of foodscapes. These principles include transparency, communication, cooperation, democracy, modularity, decoupling capacity, decentralisation, re-localisation, ecological design, adapted size, cognitive diversity, graduated commitments, and belonging. While I cannot claim to address all of these principles I argue for constant vigilance in the design of social structures towards resilience in terms of eliminating boundaries in the form of xenophobia, ageism, sexism and racism among other tensions that 'divide and constitute wage hierarchies rather than connect across commons' (De Angelis, 2017, p. 348). I conclude by suggesting approaches, including critical pedagogies of place and participatory ecologies, that can contribute to the opening-up of inclusive spaces for the co-creation of foodscapes.

Becoming food citizens

Unlike many critics of the consumer society Argentinian anthropologist Néstor García Canclini regards consumers not as 'cultural dupes'; nor does he subscribe to the view that consumer choice is a 'viable politics' (translator's introduction, Canclini, 2001, p. x). His recognition of the hybridity of tradition, the market and mass culture problematises the simple binaries that are frequently presented in discussion of food systems – alternative/conventional, community/commercial, even purpose/profit. In *Consumers and Citizens* (2001) Canclini presents identity as a 'co-production' using the case of the Mexican fiesta as a form of ceremonial 'readaptation' that reasserts and updates 'the representations of its hardships and inequalities as well as its historical cohesion' (ibid., p. xxxv). Food and music are at the centre of this state-sponsored occasion which represents, for Canclini, the nexus of the consumer, culture and tourism industries. In the context of such complex hybrid identities, he reconceptualises citizenship based on the theory that

> questions specific to citizenship, such as how we inform ourselves and who represents our interests, are answered more often than not through private consumption of commodities and media offerings than through the abstract rules of democracy or through participation in discredited political organisations.
>
> (Ibid., p. 5)

Through this lens consumption becomes more than a 'mere setting for useless expenditures and irrational impulses, but a site that is good for thinking, where a good part of economic, sociopolitical and psychological rationality is organized in all societies' (ibid.). Canclini argues that a reconstituted public sphere which bases citizenship on consumption creates 'interpretive communities of consumers, that is, ensembles of people who share tastes and interpretive pacts in relation to certain commodities (e.g. gastronomy, sports, music) that provide the basis for shared identities' (ibid., p. 159).

This may seem a peculiar starting point for a discussion of democratic participation in foodscapes however 'democracy is much more complicated than the wishful dreams of liberal theorists' (translator's introduction, Canclini, 2001, p. xxix). In a neo-liberal critique of the macro conditions influencing our foodscapes, the key dimensions of a 'strong' democracy – engagement, participation and empowerment of citizens through direct decision-making on a community level (Barber, 2004) – are difficult to achieve. Ways forward involve collaborating towards a sustainable food system through the building of coalitions; developing broad knowledge through access to information about the impacts of systems; sharing ideas or 'common talk'; developing efficacy or the capacity to determine not only one's food choices but the way foodways operate in one's environment; and maintaining an 'orientation to the common good' that recognises the value of 'mutual support and interdependence' in community food environments (Booth & Coveney, 2015, pp. 18–19).

Forms of *deeper* democracy 'draw directly upon the voices of the most marginalised and least powerful' and 'truly integrate them into decision-making processes and policies' (Carlson & Chappell, 2015, p. 3). They go far beyond political consumerism (Micheletti, 2003) and issue campaigns and require a cultural understanding of the role our food procurement and consumption habits play in citizenship. Alternative food networks (AFNs) attempt to manifest this deeper democracy. Underpinned by a strong narrative 'linking the production and consumption of local organic food to positive economic, environmental, and social changes' (Alkon & Agyeman, 2011, p. 2), AFNs propose to reconnect with and reconstitute local human, cultural and land ecologies as means to create and connect new spaces and models for engaging publics in debates over environmental sustainability, social justice and economic viability (Gottlieb & Fisher, 1996). As new forms of 'food citizenship' (Renting et al., 2012) they seek to reshape the relations between food practices and markets beyond material exchange to a 'moralisation' of food economies within which supply chains should include people of diverse backgrounds and characteristics to respect everyone's food choices and values. Yet exclusions from local agricultural activities, such as community gardens and urban farms on the basis of race, age, access limitations and real or perceived socio-economic status are common (Meenar & Hoover, 2012). The community food movement in countries including the US has been described as 'a predominately white, middle-class social change effort' (Slocum 2007, cited in Johnston & Baumann, 2010, p. 17). Garden organisers in Philadelphia, for example, a city with a history of systemic racism, reported to researchers that urban agriculture in that city is primarily a ' "White top-down" movement that is run by young White people, unintentionally excluding a non-White population' (ibid., p. 152). Some claim that immigrants and some African American communities are primarily responsible for the movement but their 'efforts and contributions are not as visible as that of White populations' (ibid., p. 153; see also Reynolds & Cohen, 2016). The food movement is also critiqued for ignoring or even rejecting the interests and cultural histories of diverse populations when establishing what constitutes 'good food' (Moore & Swisher, 2015, p. 115) and white privilege is reflected in the composition of the food movement membership, and especially its leadership (Prody, 2013). Michael Pollan was widely condemned for his list of 'food rules' intended to guide consumers towards eating practices. These included advice not to 'eat anything your great-grandmother wouldn't recognise as food' (Pollan, 2008). This neglects that 'our' great-grandmothers come from a wide variety of social and economic contexts that may have informed their perceptions of food quite differently. Pollan failed to consider the effects of race on food access and the alternative meanings his words may hold for people of colour (Alkon & Agyeman, 2011, p. 3). Culturally and Linguistically Diverse (CALD) farmers report feeling excluded from elite farmers' markets where they lack the cultural capital to participate. Literacy in English, specific cultural knowledge and familiarity with social norms are required to take advantage of the opportunities offered by AFNs. Many farmers from migrant backgrounds and especially those with limited English demonstrate a noted disinclination to engage with outside structures and organisations such as lobbies and growers associations that might assist them (James, 2016). Language is also a barrier to participation of these farmers in online forms of food-sharing (Davies & Legg, 2018).

To overcome the barriers that marginalised groups encounter in participating in AFNs Frances Moore-Lappe and Paul Du Bois (1994) offer 'ten practical arts':

1. Active listening to understand *why* people feel unable to access AFNs
2. Creating conflict to challenge the assumptions of those who control AFNs regarding issues of access
3. Mediation to bring parties together in dialogue
4. Negotiation to enlarge the space for debate and problem-solving to meet mutual interests
5. Political imagination to 're-imagine futures' according to shared values
6. Public dialogue through public forums to discuss shared issues and concerns
7. Public judgement of practices, incorporating the views of the wider community
8. Celebration as a way of expressing and communicating gratitude for what is learned and achieved
9. Evaluation and reflection to assess and apply what is learned through action
10. Mentoring to provide support and guidance for those learning these 'arts'

These steps reflect Hassanein's suggestion that food democracy 'is not only a goal or end point' but a series of processes which 'are critical to our understanding of it' (2008, p. 289). Incorporating modes of citizenship into AFN design and governance acknowledges the role of food as a 'mediator of relations within social networks' (Grasseni, 2017). Further, shifting to the view of eater as *co-producer*, rather than consumer, emphasises the reclaiming of kitchen skills, sourcing of local foods and providing fresh food access (Gottlieb & Joshi, 2010). Ideally it builds deeper connections with producers, workers and suppliers of food and stronger connections to place.

Expanding the role of the eater from consumer to that of an individual actor with practical agency demonstrates a 'more-than-food' approach (Goodman, 2016) that is highly relevant in the context of austerity and the evolving discourse of food justice. Food justice, which captures 'the struggle against racism, exploitation and oppression taking place within the food system' (Hislop, 2014, p. 19) underpins an 'action agenda' (Gottlieb & Joshi, 2010, p. 232) with 'crossover appeal' to wider social justice movements. This suggests that our ability to change our food environments is limited unless we also enlist for battle against the unjust and immensely powerful economic, social, and political forces that shape our foodscapes. Importantly, it brings into focus the social, economic and political dimensions of food choices and the 'cultural and media grammars' (Goodman, 2016, p. 264) that dictate our diets through powerful identifiers such as 'local', 'alternative', 'healthy' and 'industrial' food.

Doing local food, for example, is about reconnection with physical spaces of food production. It is also about understanding the *what* and *why* of local food in communities. Valuing local foodscapes includes acknowledgement of how food enterprises strengthen local economies and emphasise self-sufficiency; how they contribute to community-building and collective action; and their potential to promote collective ownership, social inclusion, community empowerment and collective decision-making around a range of issues such as sustainable consumption. Local food may be interpreted as a geographical concept related to the distance between food producers and consumer but can also be defined in terms of social and supply chain characteristics such as sustainable production, environmentally friendly distribution practices and social connections (Thompson et al., 2008). A helpful notion is that of the *community food system*, defined by the University of California Sustainable Agriculture Research and Education Program as 'a collaborative effort to build more locally-based, self-reliant food economies—one in which sustainable food production, processing, distribution and consumption [are] integrated to enhance the economic, environmental, and local health of a particular place' (cited in Feenstra, 2002, p. 100).

This is not to say that there is something inherently positive about local foodways. In fact, the tendency to assume this – to get caught in the 'local trap' (Born & Purcell, 2006) – has been comprehensively critiqued. Local farms and businesses cannot automatically be equated to sustainable practices and ethical outcomes (Hinrichs & Allen, 2008) and not all local provisioning systems count social justice among their aims. Further, an overt focus on localism might disadvantage diverse food-based community economic development initiatives, Farm-to-School programmes, farmers' markets, and other innovative alternative food and food justice programmes that don't adhere to strict geographical limits for procurement. Even though most local food advocates seek to identify both the benefits of eating locally produced food and the negative environmental and social consequences of an export-oriented, industrial agriculture system, Alison Alkon and Julian Agyeman's (2011) research into the 100 miles diet reveals local preference to primarily benefit those who can afford such a diet. A strategic focus on 'selective patronage' (urging consumers to purchase their food from specifically local producers) can have the unintended effect of marginalising vulnerable food and agricultural workers in distant locales at the same time that it seeks to protect and support local growers (Hinrichs & Allen, 2008). Scale is not ontologically given but socially constructed and 'an arrangement in which resources or decisions are controlled locally is no more likely to lead to ecologically sustainable or socially just outcomes than an arrangement in which the regional, national or global scale predominates' (Purcell & Brown, 2005, p. 279).

Embedded in our social networks, foodways are not only eating habits and culinary practices; they are networks of communication involving symbolic language, cultural constraints and relationships of power. They operate in non-linear patterns, including feedback and emergence that bridge the biological and social domains. Cultures of food emerge through multiple feedback loops through which our socially acquired values and beliefs dictate our consumptive habits. These feedback loops modify, reinforce and sustain our understandings about food. The system of meanings or semantic structures that generate the rules of 'doing food' are embodied materially in spoken, written and digital texts. Recognition and recalibration of world views concerning food entails interrogating the ways social structures, histories and traditions constrain our food behaviours. We integrate our experiences of growing, purchasing and eating food with ideological understandings often distorted by unequal power relations. Consuming fair trade and other ethical brands seems like an 'achievable gesture' but 'participating in foodie culture not only is a tremendous privilege, reliant on the possession of adequate economic and cultural capital, but also represents a kind of cultural hegemony' (Johnston & Baumann, 2010, p. xviii) where sense of distinction (Bourdieu, 1984) is communicated through definitions of quality, rarity, locality, organic, homemade, or simple. Identifying and incorporating different traditions of knowledge is essential to uncovering and recovering the connections that underpin the meanings we attach to food. This respects the inherent 'locatedness' of the eating experience, in that our engagement with our foodscapes is a series of 'intensely localised food events that require real choices by real people'

(Belasco, 2012, p. 3). In this sense the movement for a fairer food system is part of the wider struggle in defence of a *commons of reproduction* that includes our shared needs – food, care, energy, housing, education, water and land. Reaching consensus with one another across these 'connected subjectivities' (De Angelis, 2017, p. 387) mobilises our collective engagement in a commons movement.

Power in movement

Movement building is political action. It is essential to the construction of a new ontological foundation of community capable of navigating the challenges we face. This process is intimately tied with reconstructing political understanding to view spheres of responsibility and the role of institutions, and individuals within them, in new ways. Political participation is itself a form of reflexivity that acknowledges that we live in times of social, political and economic flux. Citizen trust in institutions of representative government, financial institutions, and the political class as a whole is low. The decline of the public sphere (Habermas, 1989), driven by the corporatisation of government and exacerbated by the failure of the mass media, is reflected in a growing dissatisfaction with political leaders. Formal politics in Western democracies such as Australia has devolved to party point-scoring and, largely through the devolution of the US as a coherent interlocutor in world affairs, social media jibes have replaced deliberative debate and diplomacy. Further, in representative democracy 'people do disagree about what they want government to accomplish, and often are obliged to act before they have the luxury of defining their preferences precisely' (Kingdon, 2010, p. 85 cited in Chappell, 2018, p. 106). In this context the real issues central to our lives – such as what we eat, who produces it and who benefits – are sidelined. We seek new forums for rational-critical debate (for a discussion see Salter, 2003).

The vital communicative dimension of a healthy public sphere is 'deliberation based on communicative rationality' or an exchange of opinions based on reasons (Habermas, 1989). Despite the fact that food has gone viral, from campaigns against genetically modified organisms to local, sustainable, 'farm-to table' dining, the fragmentation of the public sphere challenges the possibility of a common, meaningful, and effective conversation on problems of food systems. High levels of engagement are required to grasp the import of globalised political problems like food security and climate change. Environmental and social justice issues are complicated by a wide variety of standpoints stemming from identity struggles, solidarities, networks, lifestyles and imagined communities, each seeking to convince others that a particular issue is an important political problem.

Rethinking democracy as a mode of social organisation is about recognition as political voice. This theme is reflected in participative-deliberative models that go beyond institutions of democracy and the mass-mediated public sphere to consider how the marginalised or 'all affected', the weakest groups, can be included and empowered (Fraser, 2008; Callon et al., 2009). John Dewey interprets democracy as 'social cooperation' and a 'social ideal' as opposed to the more formal Habermasian model of deliberation and speech (Couldry, 2010, p. 8; Honneth, 1995). Dewey's theory is based on the notion of the 'achieved good of acting together' where democratic institutions become the 'medium through which society attempts to process and solve its problems' (cited in Couldry, 2010, p. 68). Alex Honneth (2007) notes that this is an understanding of democracy not only as a political concept but a social one. It expands the definition of democracy as an open-ended process that is 'the outcome of the experience that all members of society could have if they related to one another cooperatively through a just organisation of the division of labour' (ibid., p. 235). Central to this is caring for and recognising everybody as moral agents with unique skills and capabilities that 'are of constructive value to a concrete community' (ibid., pp. 138–139). Nancy Fraser further develops this line of thought with her argument that recognition is grounded in 'establishing a subordinated party as a full partner in social life, able to act with others as a peer' (Fraser, 2000, pp. 100–114). Couldry (2010) connects Dewey, Honneth and Fraser in claiming there is a 'direct relationship' between democracy and how we are fully realised as human beings. He references Raymond Williams' contention in *The Long Revolution* that if the individual is 'essentially a learning, creating and communicating being, the only social organisation adequate to his [sic] nature is a participating democracy, in which all of us, as unique individuals, learn, communicate and control' (Williams, 1961, p. 118).

How do we create the conditions for democratic participation? Seyla Benhabib (1996) describes deliberative democracy as a process of all citizens coming together as a public to discuss and debate common problems in free and open dialogue. Reasons and assertions are challenged; the force of the better argument decides the outcomes. Reconceptualising democracy as open participation broadens the concept of democracy as 'fundamental not only in parliaments, but also in civil society organisations: from parties to social movements, from working places to neighbourhoods' (della Porta, 2012, p. 41). These might represent 'hybrid forums' – public spaces open to all where 'questions and problems taken up are addressed at different levels in a variety of domains' (Callon et al., 2009, p. 18) by a diverse range of experts and lay persons who reach 'provisional understanding' (McFarlane, 2011a, p. 370). They are experimental in nature and appropriate to contexts where unequal power operates and legitimacy is questioned. Such spaces of participation can potentially become 'schools of democracy' (della Porta, 2013, p. 41); the more citizens participate in decision-making processes the more they are informed. Participation as such is a 'virtuous circle: opportunities to participate stimulate trust and activism, thus reproducing the stimulus to participate and improving the effects of participation itself' (ibid.). This sounds great on paper; how does it work in practice?

The paradox of participation

Heroic claims of participatory processes, both formal and informal, echo across modern democracies but outcomes include power-sharing, co-optation, and the alienation of some stakeholders, particularly when they hold diverse agendas. This is unsurprising given the connection between the economy and the environment presented in Chapter 2. In making both humans and non-humans into 'resources for investment' neo-liberalism has 'inspired investors to imbue both people and things with alienation, that is, the ability to stand alone, as if the entanglements of living did not matter' (Tsing, 2015, p. 5). Through alienation, they have become 'mobile assets'; like commodified foodstuffs we can be uprooted from our life worlds and relocated to serve the whims of capital.

As a key element of neo-liberal development reform, participation is often blamed for the failure of programmes despite being normalised in organisations and positioned centrally in digital consumption and global circuits of capitalism. Institutional design literature often focuses on meta-analyses of participatory reform but '...participation is slippery, extremely difficult to define, and any project of considerable debate has rarely been unpacked as itself a perennial, historically dynamic aspect of democratic culture' (Lee, 2015, p. 287). Ambiguous promises of participation can be both empowering and disempowering at same time. Critical urban and development studies literature shows participation is characterised by a neo-liberal ethic of 'intense possessive individualism' which has contributed to a 'political withdrawal from collective forms of action' (Harvey, 2008, p. 32). Ash Amin (2006, pp. 1018–1019) argues that the 'politics of plan' will continue to dominate the 'politics of engagement' unless governments get 'confidence in the creative powers of disagreement and dissent, in the legitimacy that flows from popular involvement'. Fundamental power differentials between scientific and governing elites and local communities constrain equitable decision-making (Armitage & Plummer, 2010). Ordinary citizens apply cultural rationality that gives equal or greater weight to personal, lived experience than 'depersonalised technical calculations' (Fisher, 2000). 'Expert' and 'ordinary' knowledges are destined to collide. The knowledge cultures of local communities can lack credibility in comparison to elite discourses (Brown, 2007). Therefore bringing expert knowledge to bear in an open and deliberative way, and bringing in 'non-expert' knowledges, challenges policy elites whose technocratic and bureaucratic decision-making has traditionally occurred with little or no input from the general public (Fisher, 2000). Valerie Brown (2008, p. 2) makes the case for synthesis-based thinking which includes 'whole-of-community engagement, collaborative inquiry and integrative management' – concepts which counter the Western ideals of 'analysis, competition, individualism and hierarchical power systems'. She argues that addressing the wicked problems we face today means treating them as 'open-ended and connected systems, rather than reduc[ing] them to single factors and fixed hierarchies' (ibid., p. 3). Yet we must maintain the advantages of specialised enquiry, and recognise that contestation and conflict are reflections of community. Each of the six knowledge conceptions in Western democracies Brown identifies strongly defend their notion of truth. All these knowledges – individual, local, specialised, strategic, holistic, nested and net – must, according to Brown (ibid., p. 31), be brought together. How to do this is a challenge of governance to achieve true public participation.

Public participation is defined as 'any process that involves the public in problem-solving or decision-making and uses public input to make decisions' (IAPP, 2016). A broader definition describes a set of conditions for participation including 'formal and informal processes and networks through which communities make decisions and attempt to solve problems' (cited in Holley, 2016, p. 10). This makes clear that civic engagement is a set of practices and conditions, including socio-economic, that provide a context for community interaction. According to Holley (2016), principles of civic engagement include access to opportunities, and justice, but these goals are problematised by unequal power relations and conflict between our diverse ways of knowing the world. Working in complex political climates where industry and corporate voices are strong is particularly challenging for FPCs, for example, dealing with controversial issues such as farm worker rights, factory farming, and school procurement systems. In their comprehensive analysis of FPCs in the US, Harper et al. (2009, p. 37) cite examples including the East New York Food Policy Council experience where some members were 'disappointed' about the lack of purity of the Council's agenda i.e. 'anti-GMO, vegetarian, pure organic'. In other cases such as the San Mateo Food System Alliance there was a 'lack of representation from youth and underserved communities' in part due to lack of resources including childcare and transportation (ibid.).

These and other dimensions of social inequality lead to polarised and uncertain civic environments that prevent the free exchange of ideas, leading to unjust outcomes that perpetuate existing injustices. Holley identifies the need for a 'big shift' in the language, structure and intent of civic engagement in terms of a 'transformative grounding in equitable and inclusive principles' that revisits the 'history, customs, assumptions and structures of relationships we have as community members, with each other and with our local community institutions … [and] commits to building relationships characterised by empathy, mutuality, and a commitment to resident and stakeholder empowerment' (Holley, 2016, p. 10). He states six principles of civic engagement: embracing the gifts of diversity; realising the role of race, power, and injustice; radical hospitality; invitation and listening; trust-building and commitment; honouring dissent and embracing protest; and adaptability to community change (ibid., p. 23).

A useful addition to this list might be adopting the 'dwelling perspective'. Tim Ingold (2000; see also Heidegger, 1971) refers to dwelling is a learned practice. Our worlds are made 'within the current' of our 'life activities' and meaning is located in the relational contexts of our 'practical engagement' with our lived environments (Ingold, 2000, p. 168). This means we learn through completing everyday tasks that require us to notice and negotiate changes in our environments, what Ingold, drawing on Gibson (1979), calls an 'education of attention' (Ingold, 2000, pp. 166–167). Tactical learning refers to the diverse modes of knowing the environment 'involved in negotiating the always contingent gap between "marginal" and "mainstream"' (McFarlane, 2011a, p. 365). A tactic is a moment in time that we seize; a 'utilisation of time' that de Certeau (1984, p. 37) refers to as a 'calculated action determined by the absence of a proper locus'. It is action taken when one is marginalised and includes a range of everyday activities including purchasing food and ways people use practical knowledge of their environment to survive. Hansen and Verkaaik (2008, pp. 12–13) extend this to the notion of 'urban infra-power' – enduring assemblages of resources and connections including 'sensing' – literally, sights, sounds, smells – and 'knowing' or decoding – understanding places of risk and danger and ways of managing these. Infra-power refers therefore to the superior knowledge of local residents who are familiar with not only 'who is who' in the neighbourhood but also with urban registers or ways of speaking, acting and gesturing that facilitate meaning-making between individuals and groups. Improvisation and adaptation are skills that develop with tactical learning within assemblages of complex relations. These skills are the product of exclusion, marginalisation and, often, exploitation. Understanding these ways of learning is vital to including tactical learners in planning processes which, in the case of food systems, must consider the entire supply chain.

Including all bodies

The extension of the food movement into the realm of social justice opens up discourse on race, class and gender in food systems. The treatment of bodies along all stages of the supply chain is being addressed in worker's rights campaigns in the US, notably the Coalition of Immokalee Workers (CIW) and United Farm Workers (UFW) in the US (Minkoff-Zern, 2017). Bodies are cheap in industrial agriculture, where pesticide poisoning, slave labour, wage theft and sexual assault persist into this century. Shifts in hiring practices including contracting, and the consolidation of the produce industry and grocery-store chains have progressively stripped voice, and power, from food workers such as the CIW's tomato pickers. The reduction of collective bargaining power through the disabling of trade unions has contributed to a growing emphasis on consumer lobbying through campaigns such as the CIW's Fair Food Campaign which asks food companies to spend a 'penny a pound' more on tomatoes and thus provide humane working conditions, including shade and water, health benefits and protection from sexual assault (ibid.).

Just as some bodies are privileged above others, so are some voices. Iris Marion Young's (1996) theory of communicative democracy refers to the 'internalised sense of the right one has to speak or not to speak' and 'the devaluation of some people's style of speech and the elevation of others' (p. 122). Usually, it is educated citizens from privileged backgrounds who have an internalised right to speak, and who have their style of speech elevated, which causes others – namely non-educated citizens from disadvantaged backgrounds – to feel intimidated. Accordingly, deliberative models of communicative democracy acknowledge social difference and the way power 'enters speech itself' (ibid., p. 123). This inclusive model acknowledges an argument can be expressed in many different ways, and in different communicative forms. It also accepts that 'people's ideas about political questions often change when they interact with other people's ideas and experiences' (ibid., p. 125). Listening is privileged by Young for its transformative potential. Listening reveals experience as perspective, provides new understandings and bridges situated knowledge. This is social learning.

Young also emphasises the importance of *greeting*, *rhetoric* and *storytelling*. Greeting is essential to democratic exchange through 'communicative gestures' by which we recognise one another and include especially 'those with whom [we] differ in opinion, interest or social location' (Young, 2000, p. 61). Narrative is a central element of self-understanding and, for Ricoeur (1980), the way we express the temporal dimensions of existence. But he also acknowledges the 'fragility of narrative identity', and how it can, when entwined with the stories of others and our own subjectivities, result in 'an unstable mixture of fabulation and actual experience' (Ricoeur, 1992, p. 162). Lives are not, after all, 'coherent plot[s]' (Couldry, 2010, p. 99). Our sense of self is not in the 'product' of the stories we tell but the 'narrating impulse' which reveals 'a narratable self with a unique story' (Cavarero, 2000, p. 35).

These perspectives provide ways of 'speaking across difference in the absence of significant shared understandings' (ibid., p. 129) and recognising the barriers to speaking and listening (Gilligan, 1982). Marginalised people are often less willing to produce and share individual narratives, unlike those who are 'the product of privilege, who can occupy the economic and cultural conditions which enable them to do work on the self' (Skeggs, 1997, p. 163). According to Young (2000, pp. 55–56) norms of speech have excluded entire social groups such as women from 'central conversations of their time'. Butler refers to the pre-discursive nature of gender (1993, p. 8) – 'the constitutive force of exclusion, erasure, violent foreclosure, abjection'. Women differ from men in their sense of moral responsibility, ethics of care and insight that self and other are 'interdependent'. But they frequently have difficulty in recognising their moral agency.

People need to be visible to be heard. The long-term exclusion of women, migrants (West, 1993), the aged (Blythe, 1979; Hazan, 1994), the First Nations people (Pascoe, 2014) and many other groups from narratives of national politics and culture reminds us that spaces for voices are inherently places of power. They are 'limited by the histories of the spaces where we find ourselves: this histories of others' struggles of recognition before us, the history of our own struggle to be recognised by contrast to particular others' (Couldry, 2010, p. 130).

Social exclusion, which Vinson (2009, p. 7) describes as 'a lack of connectedness', implicates the social and physical environments where people live. It is indicated not only by income but also social stigma on the basis of race or age and a lack of opportunity to voice one's narrative. Characteristics of social exclusion include 'limited support networks, inability to access the labour market, alienation from society and poorer educational outcomes'- all of which compound disenfranchisement. In today's Internet age, it can also be said to include digital exclusion which primarily affects individuals experiencing a lack of digital skills, barriers to connective infrastructure and limited access to technology. These material factors are complicated by affective factors such as perceptions of relevance, motivation, confidence and concerns about safety (Thomas et al., 2018).

Major traumas and losses can close off narratives completely as in cases of dispossession and violence experienced by First Nations peoples. Critical Indigenous Pedagogy of Place (CIPP) offers a way to put place and genealogy in the centre of knowledge exchange and production. Acknowledging that many epistemological approaches (how do we know what we know) do not recognise cultural knowledge, values, history and voices of First Nations people, and are therefore 'limited in meeting needs of communities or addressing structural inequalities and feelings of displacement', Trinidad (2012) argues that CIPP provides ways for people to possess agency 'not only through their roles and action but through their interpretations of their own position'. Asking 'Who is silenced? Who participates?' and ultimately 'What knowledge can best serve our community?', CIPP recognises the values and practices of a people are found in a place. It challenges positivist notions of knowledge and, when coupled with an anti-colonial stance, also challenges neo-liberalism's privileging of individual achievement and competition over the collective good and common interests of the wider community while also revalorising core First Nations values such as the spiritual connection with the land. In terms of food systems transformation, a CIPP approach demands we also enquire as to:

> *Who is the program or project really for?*
> *Who participates in shaping it?*
> *How can it be sustained?*

Responses should be based on evidence considered valid by residents in that it is local, contingent and accountable to the community. Two parallel processes exist in reclaiming First Nations epistemology: Indigenisation – putting indigenous knowledge, wisdom and concerns at the centre of practice – and re-inhabitation – learning to live well, socially and ecologically in a place that has experienced disruption and violence (ibid.).

Learning citizenship

Social learning is a process of becoming and, as such, a key element of the production of identity. Learning is not simply the acquisition of skills and information; it is becoming one who knows, and can negotiate one's way, in a specific community. While he insisted that 'all modes of human association' must embody democracy as a social ideal, John Dewey (1916) marks out educational institutions as special democratic spheres for the exercise of voice. This belief is reflected in the teachings of Brazilian educator Paulo Freire whose influence on the Latin American pedagogy of agroecology is analysed in Chapter 5.

Another fruitful concept of social learning is found in Communities of Practice (CoP) (Wenger, 1998; Lave & Wenger, 1991). These social learning systems focus on 'informal and situated social interaction, rather than on a planned mechanistic process of cognitive transmission ... achiev[ing] authentic, motivated learning of what is needed to be known about the complexities of real practice' (Cox, 2005, p. 4). They exhibit characteristics of systems thinking including emergent structure, complex relationships, self-organisation, dynamism, negotiation of identity and cultural meaning, and also locate learning in the relational space between the person and the world.

A social history of learning creates a 'regime of competence' to which participants aspire. This includes understanding the enterprise of the community and understanding what matters to that community, feeling able and equipped to engage productively with fellow members of the community and being able to access and apply the 'repertoire of resources' that the community has accumulated through its own history of learning (Wenger, 1998). Participation or active involvement in the practice is a key process in learning. Engagement provides the direct experience of competence, while imagination enables us to construct an image of a world of which we are part. Imagining is a relation of identification and is assisted by alignment in terms of coordinating our perspective, interpretations, actions in line with those of others, creating a sense of belonging. Identity is a trajectory and a 'nexus of multi-membership' – we belong to multiple communities at any time (ibid.). These identities are multi-scalar, including the home, the neighbourhood, occupation, nation, race and gender.

Importantly the concept of regime of competence connects to that of regime of accountability which, in the case of rural communities, often translates to care and stewardship of the land. As Massey (2018), writes, collective learning among a group of regenerative farmers builds a community of practice. Australia's regenerative farmers, he says, are at the forefront of an 'underground insurgency' and engaging in new emergent thinking that amalgamates organic elements with new knowledge (ibid.). He points to CoPs such as holistic grazing organisation Holistic Management and Resource Consulting Services (RSC) and the biodynamic movement as social learning systems. The farmers involved are developing an 'emergent mind' (ibid., p. 416) more focused on agroecology than agriculture, and have deep beliefs in local communities and the needs of rural and urban dwellers for healthy food. Many sell their goods directly through markets and are developing their own food brands. The centralised diffusion of farmers' adaptations that create niche markets has been reported in studies of apple farmers in Japan (Fujisawa et al., 2015), stone fruit farmers in the UK (Tompkins et al., 2010) and family farmers engaged in direct marketing in Europe (Darnhofer, 2010). These are all good examples of the alignment of institutionally-driven or 'top-down' adaptations with those of laypeople. This exemplifies the diffusion of centralised innovation (Rogers, 2004) leading to more resilient innovations (Folke, 2006) through cleverly deployed civic capacity – that is, 'the capacity to devise, decide and act collectively to improve our lives' (de Souza Briggs, 2008, p. 12).

Participatory cultures

Core structures for creating and deploying civic capacity include stable coalitions and alliances that balance the logic of empowerment (political relationships and access to influence) and the logic of efficiency (measurable progress on social conditions). These structures do not simply emerge from nowhere – they require 'deliberate strategizing ... multidimensional forms of accountability, and more practiced, skilful combinations of learning and bargaining by civic actors' (ibid.). Civil society organisations carry out important roles as brokers or intermediaries who help with problem-definition and capacity building. Accordingly, civic capacity can be viewed as the *capacity* to mobilise collective action on public problems coupled with the *choice* to exercise such agency (Vidal et al., 2001). Focusing on city municipal governments is a re-scaling of food governance that resembles a 'retreat to localism' (Sonnino, 2017, p. 1) but can actually drive transformative change. Good governance enables a government to deliver services to citizens efficiently through 'a combination of transparent and accountable institutions, strong skills and competence, and a fundamental willingness to do the right thing' (cited in Baiocchi & Ganuza, 2015, p. 198). *Inclusive* and *participatory* governance takes this a step further by drawing on the energy, influence and knowledge of citizens in addressing common problems such as food insecurity.

A more 'flexible and porous approach to the re-localisation of the food system' (Sonnino, 2017, p. 1), participatory food governance engages a large, active network of collaborative residents and professionals working together. This makes for a more resilient and democratic community than one led by heroic crusaders reliant on short-term projects with restrictive budgets. In resilient communities residents suffering severe and multiple disadvantage – many of whom experience food insecurity – are more likely to recover and sustain their well-being through supportive social networks, because, as Sir Harry Burns explains in Open Works,

> wellness is created and lived by individuals every day through their ability to care for oneself and others, being able to take decisions, and having a sense of control ... [factors that] depend upon strong social networks and high levels of social connectedness.
>
> (Civic Systems Lab, 2017)

In their 'real utopias' project, Fung and Wright (2003) identify small-scale models of 'empowered participatory governance' – real-world experiments in designing democratic institutions to draw on energy, influence and knowledge of citizens to solve common problems. These include neighbourhood governance councils, habitat conservation planning and panchayat reforms in West Bengal and in India that focus on fiscal and administrative responsibilities. They are participatory in the respect that they 'rely upon the commitment and capacities of ordinary people to make sensible decisions through reasoned deliberation and empowered because they attempt to tie action to discussion' (ibid., p. 5). As partnerships across the society and state divide they enhance the efficiency and effectiveness of policy based on two synergistic principles – *complementarity* and *embeddedness* (Evans, 1996). Complementarity is the general principle that processes and services such as collective goods can be best provided by governments while others are more suitable to efforts of private enterprise. Embeddedness is about interpersonal ties between citizens and public officials as a foundation for implementation of policies and programmes. These are relationships of mutual trust and friendship that, ideally, prevent behaviours like free riding and rent seeking.

In the well-known example of participatory budgeting (PB) by the municipality of Belo Horizonte in Brazil, five conditions contributed to success: competent bureaucracy, democratic politics, rules that regulate political influence and minimise corruption, and complementarity (Chappell, 2018, p. 166). These conditions opened up public space for citizens to gain influence over budgeting decisions. The concept originated in Porto Alegre after the military coup ended in October 1985, with the Union of Neighbourhood Associations (UAMPA), like many urban movements, growing concerned over control of municipal finances. The Worker's Party (PT) promoted PB as a concrete measure to gain voice in budgeting decisions.

But they weren't the only ones to promote PB; it was acclaimed as a policy instrument internationally, gaining traction in Europe and the United States after receiving an award for best practice from the United Nations Development programme in 1997. In fact, PB was added to the 'loose toolkit of ideas for innovative good governance' (ibid., p. 188) along with Truth and Reconciliation Commissions, the solidarity economy, citizens' juries and micro credit schemes.

Tensions arose, however, as the theory was translated into practice in Porto Alegre. The original proposal from civil society was based on an associational model in which popular leaders would represent; the actually implemented model embodied the key principle of open participation of individuals. In essence, this de-emphasised the role of the UAMPA, raising questions of representation and legitimacy. Further, community participation 'does not necessarily translate into empowerment and an increasing ability to influence the decision-making process rather than merely ratifying it' (Mathur et al., 2003 cited in Duvernoy, 2018, p. 613). The revised administrative infrastructure was an example of democratisation from above as just one possible outcome of the diversity of forms of cooperation between civil society organisations and public authorities. It is worth noting that very different local or hybrid interpretations of PB exist, and researchers note that a large part of the policy's success in Belo Horizonte was its integration in an effective administration system (Baiocchi & Ganuza, 2015).

With exceptions such as the Belo Horizonte case, the history of participatory models driven by, or integrated into, top-down interventions is littered with failures. It is widely recognised that 'marginal tinkering with programmes, and minor investments in neighbourhoods, are unlikely to foster resilient communities.... In fact, many limited and targeted grant efforts do just the opposite and reinforce separation and segregation, and in some cases even destroy communities' (Zautra et al., 2010, p. 21). The failure of many community-based interventions designed to 'help people cope with problems in living' is attributed to their focus on narrow dimensions of a problem or their concentration on short-term goals such as 'election validations'. To build resilience, as defined earlier, the recovery and sustainability of individual health and well-being is central; resilient places support resilient people, and vice versa. The creation of a participatory culture demands a whole systems perspective that transgresses age, culture and class and permits 'dissent, difference and disagreement while also confronting violence and encouraging expansive solidarities' (McFarlane, 2011a, p. 368). In this context learning is appreciated as a 'distributed process' of translation across spaces that depends on the '(re)construction' of systems across different domains, and the city itself becomes a 'pluralised, multiple learning machine' (ibid., p. 363).

One case in the United Kingdom exemplifies these principles. In the East London borough of Barking and Dagenham austerity cuts presented a 'simple choice' to the municipal authority: 'do nothing and continue to cut services, or we can find new ways of delivering them' (Barking and Dagenham Growth Commission, 2015). The borough entered into a partnership with the Participatory City Foundation to develop the *Every One Every Day* project (www.weareeveryone.org/). This initiative has spawned 250 local community projects, many initiated by residents ranging from batch cooking with neighbours to a 'bee school', Street Balls, Trade Schools, pop-up workshops on composting and DIY Film Festivals. Developed following the observation that some 'innovative citizen-led local projects were achieving inclusive participation', the participatory culture approach encourages 'common denominator activities' such as fixing, trading, singing, eating, playing, sharing, cooking, learning, growing and making – 'experiences that were co-producing something tangible as a group of equal peers' (Civic Systems Lab, 2017, p. 20).

These low/no cost, low commitment and inclusive activities can be labelled 'disruptive' in that they have the capacity to 'transform food and food experiences in ways that make them accessible to as many people as possible, especially people who might otherwise be excluded from the food system as it currently exists' (Caldwell, 2018, p. 37). They support the 'anarchist or design proposition that no one person owns ideas or knowledge or solutions' and promote 'cooperative efforts of mutual support and inspiration' (ibid.). Participants can view opportunities to participate through digital platforms which collect participation and direct outcome data to measure the collective impact of project activity including environmental and social benefits such as tons of food grown locally; square metres of land under cultivation; zero carbon journeys; hours spent in physical activity; and overall sense of happiness and well-being (Participatory City Foundation, 2018). These platforms are increasingly useful in collecting and measuring the public metrics and other data required to drive and measure the success of the project. While this project is in its early stages (Year One of a five-year plan), it describes a politics of engagement that privileges learning in a place where residents experience diverse attachments, anxieties, and sometimes ambivalence about where they live.

Conclusion

As argued in this chapter participation is a form of emancipatory learning that can change attitudes, perspectives and influence beliefs. It is a means of redistributing resources, reducing inequalities and facilitating the recognition and representation of those rarely heard from in local governance. The development of participatory cultures is challenging, not least because it involves a complexity of actors. Negotiating voice and participation in complex societies is a multi-scalar project which incorporates individuals, families, local organisations, communities, polities comprising large communities, federated political units and global political organisations (Couldry, 2010, p. 100).

Embracing the challenge of scale is essential to tackling meta-political misrepresentations on a global level, which emerge 'when states and transnational elites monopolise the activity of frame-setting [the 'who' in addressing social justice on a global level] denying voice to those who may be harmed in the process' (Fraser, 2005, p. 85). The organisational challenges of scale are addressed, in theory, by democratic politics as the recognition of individual voice as collective voice. However, distributed forms of voice raise particular issues. Political action through decentralised networks is based on representative processes of formal and organisational politics where the struggle to respect the diversity of participants is played out in a variety of local, regional and international forums.

Transnational social movements for social justice are examples of these decentralised, educative and adaptive entities which draw on participants' relational capacities and cultures through global networks (Poletta, 2002). In supporting collective and social rights, they purport to be democratic, offer structures and forms of horizontal organisation that embody this principle, and promote a participatory ethos that changes discourse 'away from technocratic elite discourse [to] a new discourse that emphasise[s] popular participation' (Mainwaring, 1987, p. 149 cited in della Porta, 2013 p. 144). Social movement repertoires of action such as a 'logic of membership' favour participatory democratic models, foster opposition and provide 'arenas for plural, inclusive coalitions during phases of liberalization and transition' (della Porta, 2013, p. 146). They also allow, della Porta adds, for 'more civil society mobilisation in the successive periods' (ibid.). These processes are fraught with friction (Tsing, 2005) as participants grapple with the tension between articulating an agreed political statement that distils the views of members into clear communications without the emergence of centralisation and hierarchy (Juris, 2008). I explore these and other challenges through an analysis of the communicative constitution of the global peasant movement LVC in the following chapter, Chapter 4.

4 Organising through communication

The current political organisation of agriculture, the purpose of which is 'primarily to produce profits, incidentally to produce food' (Allen & Sachs, cited in Massey, 2018, p. 485) has triggered the emergence of transnational social movements concerned not only about what we eat but about the specific function of rural space, the type of development that should take place in it, and who should benefit from this. These movements, divergent in origin, converge in a common desire to overthrow neo-liberalism as the sole set of values. They are committed to building community and the civil commons. Although defensive in orientation, social movements 'create the conditions for offensive struggles, through changes in paradigms, redefinition of sense horizons, clarification of the nature and structure of fields of power relations, and new connections with other movements through solidarity' (De Angelis, 2017, p. 369).

The roots of local conflicts long pre-date the establishment of transnational movements such as La Vía Campesina (LVC), the case study presented in this chapter. In the case of Brazil, concerns over the harmful effects of pesticides and the damaging social impacts of modernising agriculture including concentration of land tenure and growing inequality mobilised resistance in the 1980s. The desire 'to produce food without poison' motivated projects in Serra Gaúcha and Zona da Mata with the aim of making 'concrete' a model of agriculture focused on alternative farming practices without use of agrochemicals (Charão-Marques et al., 2017, p. 129). In the case of the Vacaria Project in Serra Gaúcha agronomists sought to provide proof in debates over agrochemical laws. Their locally constructed concepts of 'ecologist famers' and 'ecologically based agriculture' were framed to make explicit their opposition to organic farming, which was understood as 'a narrow political strategy oriented to input substitution and the creation of niche markets' (ibid., p. 130). This public critique of organic certification led to the emergence of the *Rede Ecovida de Agroecologia*, a participatory certification and framework that opened space for diverse forms of sustainable food production other than those dictated by the state. The movement evolved with new pressures to intensify chemical use in response to rising coffee prices in the 1990s. This motivated the campaign "Movement in Defense of Life and the Environment" which denounced pesticides and also appealed for the right to land. 'Outbreaking associations' including seed saving networks focused on rescuing and using creole seeds while regional women's organisations linked the homogenisation of the food market, dependency on hybrid seeds sold by transnational companies and the oppression of women.

In elevating food from object to subject (Charão-Marques et al., 2017) these local initiatives pose a potent threat to conventional agriculture. This chapter aims to provide some insight into how local campaigns are translated into a flexible and inclusive global agenda by LVC. I focus on the communicative processes and interrelationships that occur in organising, with the aim of addressing the question of how a set of people, practices and messages become a *democratic* organisation through communication. First, the intersection of social movement and organisational theory – both of which seek to understand the nature of collective action – is outlined. The communicative constitution of organisation (CCO) approach is applied in analysing the processes that link farmer-activists within and across national boundaries, shape operations and situate local actors in wider contexts. I suggest that this theory can help us understand voice as a site of domination and resistance, the differential capacities and forms of engagement and participation in specific grassroots organisations, and the complexity of multidirectional processes of scaling required for global action.

Social movement as organisation

Social movements are 'dynamic communication systems' (Fuchs, 2006, p. 101) within which individual and collective communication practices produce alternative understandings and oppositional framings that contest and seek to generate 'large-scale, collective changes in the domains of state policy, corporate practice, social structure, cultural norms and daily lived experience' (Ganesh et al., 2005, p. 177). The communicative constitution of organisation (CCO) approach is based on the assumption that organisations are 'complex resource[s] with implications for individual agency' (McPhee & Iverson, 2009, p. 74) constituted in and through human communication. From this perspective organisations are not objects or social facts within which communication occurs; they are realised through communication processes. Usually applied in the analysis of firms and other more formal structures, the CCO approach is a novel theory for the study of networked social movements. It is also a very valuable theory in that it avoids conventional thinking around how movements emerge, endure, and affect change in the global polity.

Communication is a constitutive force where a receptive audience confers socially recognised agency on a speaker, making the entire communication process the 'locus of constitution' (McPhee & Zaug, 2000, para 14). Different audiences demand distinct relations with an organisation, driving the need for a variety of messages. The four-flows model of organisational constitution (McPhee & Iverson, 2009; McPhee & Zaug, 2000) emphasises circulating systems or message flows in four categories: membership negotiation, self-structuring, activity coordination and institutional positioning (Putnam & Nicotera, 2009). The model can reveal how 'the organisation as a whole manages to develop an identity (self-structuring) and, by itself becoming an actor, how it can subsequently negotiate a relationship with its members (membership negotiation in a different mode) and outsiders (positions itself institutionally)' (Taylor, 2009, p. 166). The four flows involve 'crosscurrents' that play a role in constitutive communication beyond information transmission; episodes of communication are interactive, involving multiple participants with varying agendas and goals (McPhee & Zaug, 2000). Chains of interactions and episodes diffuse upwards and outwards through the organisation through formal and informal communication processes.

These processes sustain the constant negotiation (and renegotiation) of meanings across the 'participating plurality' who engage in the work of 'singling out the commons that they are defending and why, of mapping out friends and foes, of drawing distinctions and of developing a common-sense horizon' (De Angelis, 2017, p. 369). In doing so social movement participants demonstrate communication to be 'the ongoing, dynamic, interactive process of manipulating symbols toward the creation, maintenance, destruction, and/or transformation of meanings' (Kuhn, 2012, p. 570). These meanings, and the communication processes that create and sustain them, are pivotal to the existence of the social movement as an organisation or social form created and maintained by members who respect it as a high-order system and make provision for its survival (McPhee & Zaug, 2000).

Research at the intersection of social movement and organisational studies has been prompted by the recognition that processes such as framing, collective identity, and mobilisation can be applied in understanding the political and cultural dimensions of organisations (Weber & King, 2014). The social constructivist view that organisations are rooted in and emanate from communication is underpinned by the belief that cultural change can be mapped through changes in discourse, which here is understood as 'a shared set of concepts, vocabulary, terms of reference, evaluations, associations, polarities and standards of argument connected to a coherent perspective on the world' (Rochon, 1998, p. 16). In articulating claims and mobilising claimants, transnational social movements are discursive networks engaging in 'contentious performances' (Tilly & Tarrow, 2007, p. 11) through which collective claims are made on other political actors including states and global governing bodies. These claims embody culture and history; they are situated in broader, dynamic political cultures that are comprised of values and beliefs that influence activist engagement in practices including strategising, organisation, protest and the formation of collective identity (Kavada, 2013).

The CCO approach offers a conceptual foundation from which to advance communicative explanations of social movement organising. Organisations are socially constructed from networks of ongoing conversations that maintain a particular version of 'reality' for participants. Through discourse, meanings are produced, challenged and renegotiated. The production and consumption of media texts, interpersonal and group communications – along with the topics, frames, discourses and forms of different media – combine to represent the communication culture of the movement. Patterns of communication are 'reflexively identified and retained by members' in the constitution of a 'social entity' called an organisation (McPhee & Zaug, 2000, para 3). The conceptualisation of social movement as organisational field – 'a set of interacting groups, organisations and agencies oriented around a common substantive interest' (Aldrich, cited in Minkoff & McCarthy, 2005, p. 291) is useful in explaining the mechanisms and processes through which social movement organisational forms not only emerge, attain legitimacy, adapt and survive, but are produced and reproduced continually in various power configurations (McPhee & Zaug, 2000, para 7). Social movements that endure for long periods of time, like LVC which emerged in 1993, tend to develop more formal mechanisms of coordination and are therefore more amenable to organisational analyses (Davis et al., 2005).

Where CCO theory can contribute most to social movement studies is in realising the goal of 'bridging the micro/macro gap, by showing how to discover the structure in the process and delineating the processes that realise the structure' (Taylor, 1993, p. 261). The processes of communication create a 'patterning that constitutes the structure of organisation and the organisation itself simultaneously' (McPhee & Zaug, 2000, para 6). Mapping how these processes operate across national public spheres includes determining the degree of centrality of transnational exchanges to the daily routine of locally-based member organisations.

The second valuable contribution that CCO can make to social movement analysis is in explicating configurations of power in the constituting process to offer a critical theory of how that power is manifested and shapes organisational members' reality and their capacity for voice. This is particularly relevant in studies of transnational 'movements of movements' such as LVC comprised of a diverse membership of organisations with varying degrees of influence within regional coalitions. Power is 'an inevitable and constitutive element in all social and institutional interaction [and] all communication necessarily involves the use of power' (Deetz & Mumby, 1990, p. 37). Within social movement organisations that value low levels of formalisation, structure and hierarchy, participants must remain alert to how communication produces and reproduces particular power structures to the exclusion of alternate power configurations. In periods of change social movements, like all organisations, are arenas of struggle populated by challengers from both within and without (Fligstein & McAdam, 2012).

To this end, the CCO approach enables us to more fully recognise processes of network construction and capacity building that include productive relationships in cooperation with partner organisations rather than viewing the organisation as a 'container' with 'spatial limits' within which communication processes occur (McPhee & Zaug, 2000, para 4). Within the transnational social movement network – unstable, contingent and responsive, with the 'ability to combine rule-governed organisation with flexibility, open-endedness, de-centredness and spatial dispersion' (Fraser, 2008, p. 128) – processes of self-structuring rely on reputation, cooperation, trust-building, and the clarifying of expectations without reliance on 'rigid control' (McPhee & Zaug, 2000, para 30).

Finally, the CCO approach can contribute to a subaltern perspective on organising for collective action that challenges the dominant, liberal scholarship on networks. Such scholarship is, arguably, 'limited in terms of understanding movements that have a more collective character and a style of action that goes beyond issue campaigns and policy reforms' (Escobar, 2008, p. 271). Research focused on transnational advocacy networks (TANs) interrogates network elements such as the centrality of NGOs, the acquisition of resources for policy-making, and the motivations for alliance building in these processes (Keck & Sikkink, 1998) but disregards the knowledge of the Global South and assumes that networks serve to fulfil instrumental ends (Borras et al., 2008). In reconceptualising networked social movements Arturo Escobar (2008) proposes that subaltern actor networks and dominant actor networks have contrasting goals, practices, mechanics of emergence and ways of working. He draws on the work of Mario Diani (2003, p. 1) who describes collective action as shaped by social ties between participants, 'groups and organisations, with various levels of formalisation, linked in patterns of interaction which run from the fairly centralised to the totally decentralised, from the cooperative to the explicitly hostile'. In describing networked social movements it is vital to recognise the 'fallacy of composition' which alerts us that 'you cannot infer something is true for the whole from the fact that it is true of some part' (De Angelis, 2017, p. 324). The CCO approach provides a template for studying interactions and frictions, and therefore has the potential to provide a broader perspective on processes and actors that depicts social movements as 'expansive, heterogeneous, and polycentric discursive fields of action, which extend well beyond a distinct set of civil society organisations' (Escobar, 2008, p. 272). The field of action is constantly reconstructed and shaped by changing power relationships between movement cultures where political meanings are contested by alternative, subaltern publics undergoing processes of reinvention and articulation. Creating and sustaining alternative discourses, social movement participants contest their own agendas, facilitating flexibility and responsiveness within their own principles.

The value of the CCO approach to this subaltern perspective lies not only in acknowledging that organisations occur in a variety of ways and that no one communication form can sustain them, but also in recognising that organising is primary. Actors who forge connections are important but peripheral to the act of organising itself. Connections 'construct the identities of actors' (Czarniawska, 2013, p. 13). Accordingly, CCO theory can help fill 'silences in the literature' (Borras et al., 2008) on the internal dynamics of rural social movements, issues of partial representation, processes of interconnectivity and the presence of 'trans-community' links such as those between rural, women's, and indigenous movements (Yashar, 2005). It broadens the conceptualisation of social movement ecology by contextualising the multi-perspectival politics of social movement participants while providing the scope to explore the role of movements as 'learning networks' (McMichael, 2008) within which central values, beliefs and principles are shared through a range of interactive processes.

The four flows

Process, equivalence, structure, and power underpin the typology of message flows devised by McPhee and Zaug (2000). According to their four-flow theory, organisations are constituted by four analytically distinct yet overlapping flows of interaction processes: (1) membership negotiation, in which members define their relationship to the organisation and other members and set boundaries; (2) self-structuring, in which interactions design, implement, and control organisational processes; (3) activity coordination, adapting interactions to work situations and problems; and (4) institutional positioning, situating the organisation in its larger environment. Each flow is an interactive communication episode, typically involving reproduction of, as well as resistance to, the rules and resources of the organisation. Flows control and condition one another and events are part of an ongoing stream of discursive practice within which single messages can address more than one constitutive task.

Membership negotiation

One of the processes essential to social movement organisations is the establishing and maintenance of a relationship with members. Organisations come to be when they attract members and introduce them to a world unique to the organisation. The natural medium of sense-making in organisational life is the narrative or 'meta-story' through which 'many' become 'we' (Taylor, 2009, p. 178). Narrative identifies a problem by initially establishing the facts and a common basis of understanding, what is to be done about it, who will do it, before assessing what was done and what the outcomes were. The role of 'culture, collective emotions and biography' (Weber & King, 2014, p. 13) in these processes cannot be ignored.

LVC states itself to be comprised exclusively of smallholder farmers, landless peasants, fisherfolk, migrant and seasonal workers who number over 1.2 billion, globally (Saragih, 2012). Brought into being through a series of meetings between Latin American, European and Asian farmers and sustained by farmer-to-farmer research and grassroots extension approaches, the movement embodies a form of contentious politics fuelled by 'participation-boosters' or strategies that 'endow agency' (Kröger, 2011, p. 438; see also Altieri & Toledo, 2011; Holt-Giménez et al., 2010; Rosset et al., 2011). To be a member of LVC one must be an active participant in a locally-based producer or rural workers' organisation. The collective self-representation of members is embedded in deeply ingrained cultural understandings of the relationship between the farmer and the land, embodied in the aforementioned concept of food sovereignty. The narrative of food sovereignty lends a positive distinctiveness to members, a sense of self-definition and self-representation. This binds them internationally in a common purpose, communicated in the LVC slogan, 'Globalise the Struggle, Globalise Hope', and in the mission:

> to develop solidarity and unity among small farmer organisations in order to promote gender parity and social justice in fair economic relations; the preservation of land, water, seeds and other natural resources; food sovereignty; [and] sustainable agricultural production based on small and medium-sized producers.
>
> (La Vía Campesina, 2008)

Words, architectural elements, performative devices and artefacts contribute to self-representation and membership identity. Nyéléni, a purpose-built sustainable village in Sélingué, a small town in Mali, Western Africa, was the site of the first International Food Sovereignty Forum in February 2007 and the International Forum of Agroecology in 2015. Constructed with local labour out of local materials, Nyéléni serves as a permanent communal space for African and international social movements to meet. It is a meeting place that symbolises collective rights, access to the commons, and *diálogo de saberes* – a Freirean form of farmer-to-farmer or *campesino a campesino* (CAC) knowledge exchange based on 'horizontal dialogue between peers who have different knowledges and cosmovisions' (Martínez-Torres & Rosset, 2014, p. 4) which is central to the communicative constitution of LVC. A second unifying device is *mística*, a shared ritual performance of the connection between the peasant, seeds, soil and water. As a way of creating 'a sense of cohesiveness among people from such diverse and different cultures who do not speak common languages' (ibid., p. 164) *mística* is a communicative device that fosters the development and maintenance of a common identity within the movement. It serves as a bridge between communities of practice; people who share our concerns and understand us when we talk to them (Taylor, 2009).

Cultivating and maintaining a global constituency requires LVC members to translate their localised interactions and 'situated conversations' (Koschmann, 2011, p. 66) to wider audiences. But efforts to internationalise membership are hampered by limited actual and virtual interactions between members of the network. Face-to-face meetings at conferences are irregular due to the expense of air travel. In some regions extending networks has proven to be problematic; bottlenecks can be created when the membership process places too much emphasis on national and regional levels, where dominant members tend to gate-keep (Desmarais, 2007). This tendency reflects how social movement organisations are subject to processes of power negotiation through 'authoritative meta-communication', power-claiming, and struggles to speak on behalf of the organisation (McPhee & Zaug, 2000, para 31). At their Fifth Conference in Mozambique, members formally acknowledged the importance of strong local and national organisations working at a regional level to the future of the movement:

> In the positive process of transformation of Vía Campesina into a broader and more active and more decentralised movement we have to assure that we also strengthen the consultation processes among the national organisations and regions and that we strengthen the participation of the leadership in the regional and international processes.
>
> (La Vía Campesina, 2009, p. 54)

Strengthening links in the network through enhanced communication and information exchange remains a priority as does a strong focus on gender equity, reinforced by the early establishment of a women's chapter. LVC emphasises that food sovereignty is 'only possible with a fundamental transformation of unequal gender relations within and beyond movements themselves' (Desmarais & Nicholson, 2013, p. 7). The movement's 'World Campaign to End Violence Against Women' identifies neo-liberalism with patriarchy, linking local struggles against everyday forms of dominance within the capitalist market.

It is also noteworthy that LVC's traditional pan-Latin American uniform of green cap and pañuelos or scarf, a symbol of membership worn at protests, has been supplanted by local headwear at regional conferences in South-East Asia. This acknowledges the growing footprint of the movement and its increasingly diverse membership. Such modifications indicate the social movement organisation is not an 'it' but is a becoming (Taylor, 2009, p. 182). It is an object of reflexive control and design, or self-structuring.

Organisational self-structuring

Self-structuring processes include 'any process that steers the organisation or part of it' (McPhee & Zaug, 2000, para 29). The vision of food sovereignty is explicitly invoked in LVC's discussion papers, campaigns and official declarations repeatedly and its members speak and act in the name of this vision, positioning food sovereignty as a 'figure participating in the definition of organisational situations, which is another way to speak of its agency, delineated in and through communication' (Cooren et al., 2011, p. 8). The concept is a central trope in the constitution of the organisation and the official documents that steer the organisation and the activities of members. As self-structuring processes these consensually developed understandings are crucial to building a foundation for member participation and cooperation based on 'low uncertainty, high mutual knowledge and high goal alignment' (McPhee & Zaug, 2000, para 29). These attitudes form the basis of the legitimate power generated through the democratic processes of decision-making that operate within the movement, a 'meta-layer of self-structuring process that grounds and solidifies the collaboration' (ibid., 2000, para 30). The specific communication methodology *diálogo de saberes* – as defined earlier – is central to the self-structuring processes of LVC. Farmer-driven programmes generating and sharing technologies and solutions in regular gatherings spread knowledge in a horizontal and decentralised manner that is designed to mitigate forms of oppression and hierarchy. As a place for the sharing of ideas, experiences, 'parables, stories and humours' (Altieri & Toledo, 2011, p. 600), the importance of these meetings in creating sense of unity is significant. Nevertheless, even strong supporters of the methodology agree that spreading knowledge of sustainable agricultural practices does not 'address the need for an enabling policy context' (Rosset et al., 2011, p. 206) that formulates and presents viable alternatives to global threats. The building of a wider movement is the focus of 'purposive assemblies' (Kröger, 2011) that foster the common identities, frames, and demands of local groups based on a common analysis of the crisis of agriculture under neo-liberalism. LVC's international conference is its highest decision-making body, meeting every four years to shape the general principles and campaigns of the transnational movement through participatory means. Two representatives of each of nine global regions – including at least one woman and, more recently, a youth representative – attend the meetings. Further balance is achieved through the regular rotation of the international secretariat which was relocated from Indonesia to Zimbabwe in 2013 reflecting not only the vulnerability of the African continent to global concerns, such as land grabbing and impacts of climate change, but the desire to extend the footprint of LVC in the region (Mann, 2014a).

Meetings at Nyéléni have produced important action-oriented steering documents in the form of Declarations that solidify the connections between national and regional social movements and facilitate the development of a common concept of food sovereignty. The Declaration of Nyéléni (2007) presents specific goals of peasant organisations, small-scale farmers, fisherfolk, shepherds, indigenous peoples, forest communities, women, consumers and environmentalists campaigning for food sovereignty. It also outlines strategies to achieve these goals. The Declaration of the International Forum of Agroecology (IPC, 2015) aims to further progress this political project through championing traditional peasant farming methods as an alternative to conventional industrial agriculture. Powerful statements and symbolically coded representations of jointly experienced circumstances are examples of collective sense-making whereby a 'situation' is comprehended and articulated specifically in words that serve as a 'springboard for action' (McPhee & Zaug, 2000, para 20). These Declarations serve as a blueprint for communication between members, coordination of tasks, allocation of resources and the implementation of collaborative decision-making processes. These processes are integral to the search for 'ideational pathways' (von Bülow, 2010) in the form of common agendas and collective action frames that play a central role in the communicative culture of the movement and the construction of collective identity, 'a strategic resource that organisations create through interaction and communication with members' (Bimber et al., 2012, p. 82). Thus, individual members of LVC's network experience unity through the scale-shift to transnational collective action, a process through which 'individuals, non-state groups, and/or organisations mobilise jointly around issues, goals and targets that link the domestic and international arenas' (von Bülow, 2010, p. 5). However, self-structuring processes are also subject to 'discrepancy, dispersal and ambiguity' (McPhee & Zaug, 2000, para 29) in response to specific national constraints and opportunities. This is revealed in analysis of the third communication flow, activity coordination.

Activity coordination

Analytically distinct from the self-structuring processes that determine the internal operations, decision-making, and norms of the wider movement, the activities of LVC members in specific countries are focused on the achievement of objectives based on local interpretations of food sovereignty. At a country level, activity coordination, while based on policies and plans for work agreed in international and regional meetings, adapts to specific situations and problems (McPhee & Zaug, 2000, para 22). Adjustments and modifications take place as members 'presume that they are working not just on related tasks but within a common social unit with an existence that goes beyond the work interdependence itself' (ibid., para 37). Processes that lead to practical organisational possession include budgeting, marketing campaigns, official (state-based) declarations, and meetings along with daily interactions. Group tasks lead to 'the slow sedimentation of organisational culture and knowledge' (McPhee & Iverson, 2009, p. 78) and become a 'foundation of [local] activist networks ... making information and communication practices central to the identity and existence of the group' (Breindl, 2012, p. 25). There is a clear sense of working outside the mass-mediated public sphere:

> We must continue to accelerate political, ideological, organisational and technical training based on our own concepts, since we know that conventional education suppresses our identity and our ideas. Such training is crucial for our movements to create new and committed actors to forge our own destiny. In our struggle we also need to continue to create our own autonomous communications media, and build our alliances with the alternative media, to build awareness of our culture, our dignity and our capacity to transform society.
>
> (Euskal Herria Declaration, La Vía Campesina, 2017a)

Given this sense of working on the periphery of hegemonic systems, LVC bypasses traditional media channels to focus on alternative media platforms and a highly active network of capacity-building partners such as FIAN International (the Food First Information and Action Network) and Friends of the Earth. Co-authoring and posting reports such as 'Grassroots resistance to the corporate power of agribusiness in the era of the "green economy" and a changing climate' (Zacune, 2012) and publishing empirical research to support the argument that small-scale, agroecological farming methods 'produce high yields, environmental and social benefits' (Laughton, 2017) the movement has gained credibility and legitimacy in international civil society and governance arenas such as World Committee for Food Security within the United Nations Food and Agriculture Organisation (FAO) of the United Nations, where it participates in the Civil Society Mechanism (Mann, 2014a). Drawing on the resources of news outlets including the International Consortium of Investigative Journalists, LVC hyperlinks to reports on corporate wrongdoing and failures of global governance, reframing the struggle for food sovereignty in the wider campaign for social and economic justice. Activist members also engage in storytelling via Twitter, Facebook and Instagram, platforms that foster the growing sphere of global social media activism by providing new opportunities for the articulation of concerns and dissent to wider audiences (Cottle & Lester, 2011; Mann, 2018b). The movement has effectively broadened its capacities for outreach through fluid collaborations with global justice networks such as the World Social Forum, integrative campaigning and the use of alternative media platforms, such as Indymedia.com, and has developed *La Vía Campesina TV* (https://tv.viacampesina.org/) to share rich multimedia content:

> This television is ours; it is a mirror of our daily lives, our dreams and our struggles … it brings together audio-visual material and cultural creations from the farmers' organisations themselves, whether they are members of Vía Campesina or not, as well as material produced by friends, allies or sometimes mainstream media.
>
> (La Vía Campesina, 2013b)

Described as a 'Babel Tower', the site features expert and non-expert voices in multiple languages. Engineering and enacting these communicative events and creating media objects that have cultural meaning in a network of production and reception, the movement demonstrates communication as 'the ongoing, dynamic, interactive process of manipulating symbols toward the creation, maintenance, destruction and/or transformation of meanings' (Ashcraft et al., 2009, p. 22).

The campaigns of local organisations are rooted in local realities; they reflect the multiplicity of unique challenges, attitudes and constraints experienced by individuals in different countries. The activity coordination practised by Mexico's National Business Association of Rural Producers and Traders (ANEC), for example, reflects its membership, issue agenda and the local political environment. A cooperative of grain growers, ANEC has responded to Mexico's rapid integration into the global trade system through NAFTA by maximising economies of scale. Collectively purchasing and maintaining warehousing and transport facilities, instigating rural insurance programmes and providing regular market updates to small-producers, ANEC enables its members to compete with agro-industrial complexes such as the food conglomerates Cargill, ADM (Archer Daniels Midland – 'Supermarket to the World'), and Monsanto. Activity coordination largely focuses on direct action and lobbying government in regional coalitions such as the MesoAmerican Movement of Indigenous and Peasant Farmers (MOICAM) and the Association of Central American Peasant Organisations for Cooperation and Development (ASCODE) against NAFTA and the Security and Prosperity Partnership (ASPAN). ANEC teamed up with Greenpeace México to create the campaign 'Tortillas de las buenas, tortillas de maíz mexicano, libres de transgénicos' (Good tortillas from Mexican corn, GMO free) aimed at raising awareness among Mexican consumers of the potential health risks of consuming US corn and the importance of supporting Mexican campesinos. In 2009, as part of the national campaign that organised many of the demonstrations captured by the global news media, 'Sin Maíz No Hay País' (No Corn, No Country), activists decorated the main square, the *zócalo*, with images of Mexico and campaign slogans in tonnes of maize. Other actions have included public fasts, a blockade of the cross-border bridge at Ciudad Juárez, a traffic-stopping 2,000-strong tractor convoy and rock concerts targeting young Mexicans. In January 2013 ANEC joined other rural organisations as part of the National Union of Autonomous Regional Peasant Organisations of Mexico (UNORCA) in a hunger strike at the Angel Monument in Mexico City, appealing to President Enrique Peña Nieto to prevent open-field plantings of genetically modified maize in the northern states of Sinaloa and Tamaulipas, drawing declarations of support from around the world. These transnational ties have enabled ANEC and other campesino organisations to 'go global' in 'reframe[ing] their platform in terms of a defense of global genetic commons against a new corporate enclosure movement'

(Richard, 2012, p. 71). The relationships with other civil society organisations forged through these activities demonstrate the relatedness of the four flows, as they also serve as examples of institutional positioning.

Institutional positioning

Institutional positioning is key for any organisation aiming to recruit members and gain the support of wider audiences. The 'constitutive moves' an organisation makes in establishing itself as a viable entity or force in the public sphere include connection and interaction with important elements of the political and social environment. Through these interactions the organisation aims to 'negotiate an image as a viable relational partner' (McPhee & Zaug, 2000, para 39). This has always been a primary concern of LVC, which first emerged as a direct result of the dissatisfaction of small-scale farmers with their representation by the International Federation of Agricultural Producers (IFAP).

Within national public spheres institutional positioning is constant and facilitated by journalists, public relations practitioners and lobbyists who act as brokers linking and mediating relations between previously disconnected actors (Ganesh & Stohl, 2010). The efforts of social movement organisations to engage with these brokers are influenced by perceptions and understandings about the contemporary media and political environment. In theory, the existence of a free, independent press is an element of the media environment that fosters diversity of opinion in democratic public spheres (Habermas, 1989). When the press is ideologically aligned with a position, support can be strong as illustrated by ANEC's strong editorial presence in Mexican daily newspaper *La Jornada*. Digital technologies have enabled activists to circumvent mass media organisations in forming transnational alliances and positioning themselves institutionally. This capacity, coupled with the search for alternative hospitable venues for claims-making beyond national public spheres, makes electronic communication through websites, email and listservs an increasingly important component of social movement activism, as the case of the highly networked Combat Monsanto (www.combat-monsanto.org/) campaign demonstrates. Jointly sponsored by LVC, Greenpeace, Friends of the Earth, Sherpa, ATTAC and Fondation Sciences Citoyennes, this example of digitally networked activism is uniting farmers and consumers internationally and was credited for having a major role in the development of a de facto ban on GMO food crops throughout Europe. It encourages eaters to avoid GMOs on health and cultural grounds and challenges agribusiness framing of genetically modified foods (GMFs) as safe and crucial to addressing world hunger. The reframing of GM food as 'Frankenfood', which continues to circulate as a global meme, highlights risk as a mobilising frame in the debate and exploits the scientific illiteracy of the lay public. Friends of the Earth Europe (FoEE), as the largest grassroots environmental network in the region, unites more than 30 national organisations with thousands of local groups in seeking to influence European and EU policy and raise public awareness on the environmental threat GMOs pose to biodiversity. FoEE effectively uses political platforms such as Radio Mundo Real, a multilingual web-based outlet, and ARC2020, the Agriculture and Rural Convention (www.arc2020.eu/). Initially created as a multi-NGO platform to influence the reform of the CAP between 2010 and 2013, ARC2020 now focuses on stimulating public debate in Europe on future farming, food and rural policies through alliance building and communication on a variety of platforms. They host debates on Facebook and Twitter (#afterCAP, #milkcrisis, #livestockdebate), and post rich, shareable video content that promotes food sovereignty such as the FoEE briefing 'Feeding Europe' (www.arc2020.eu/feeding-europe-food-sovereignty-and-agro-ecology/).

Institutional positioning has been further enabled through the facilitation of increased engagement with the FAO. In global food and agriculture policy forums LVC has gained considerable standing since the food price crisis of 2007–08, particularly via the UN system (Mann, 2011, 2014a). The work of the UN Special Rapporteurs and capacity-building organisations such as FIAN International has been instrumental in progressing the food sovereignty agenda through human rights frameworks, promoting policy instruments, such as *The Declaration of Peasant Rights – Women and Men* (2018) on the grounds that peasants face specific threats to their rights, being among the most discriminated against and vulnerable people in the world, and the main victims of poverty and hunger. The adoption of this Declaration on 28 September 2018 (OHCHR, 2018) follows the previous acknowledgement of UN FAO Director-General José Graziano da Silva that support to smallholder farmers is essential, claiming 'we are making up for lost time, as food security governance was neglected until a few years ago' (da Silva, 2012; Golay, 2010). Civil society actors such as LVC have secured more meaningful involvement in planning, research, debate, and policy-making about food and agriculture through strong partnerships with the International Planning Committee for Food Sovereignty (IPC), an autonomous and self-organised global platform of smallholder farmers and rural workers' organisations aiming to progress the food sovereignty agenda on global and regional levels. Their efforts have led to widening participation in international governance after decades of marginalisation in discussions over world food security. The inclusion of civil society organisations as official participants in the Committee on World Food Security (CFS) in late 2009 has been a landmark achievement in securing policy engagement for social movements. Established in the 1970s in response to an earlier food crisis, the original CFS was largely ineffective in meeting its mandate to review and follow up food production and security issues, and was poorly supported by member states in terms of commitment and funding (Shaw, 2007). The reformed CFS is promoted as a new space for engagement, a multi-stakeholder forum committed to eliminating hunger following three guiding principles: *inclusivity*, to ensure that voices of all relevant stakeholders are heard in the policy debates on food, agriculture, and nutrition; *strong linkages to the field*, to ensure the work of the CFS is based on the reality on the ground; and *flexibility*, in the face of a changing external environment and the needs of countries (CFS, 2012, p. 9). Unlike the case in previous UN platforms, civil society actors are recognised as meaningful political agents representing their own interests in the CFS mission to develop a global strategic framework on ending hunger. An autonomous international food security and nutrition civil society mechanism (CSM) facilitates the participation of organisations from countries that regularly experience food insecurity, many of which are members of LVC. Five years on Nora McKeon (2015, pp. 184–185) described the CSM as an 'innovative interface' that represents a 'brave new experiment in civil society self-governance'. Problematic issues, particularly for poorly resourced peoples' movements, include the dominance of the English language, insufficient time and human resources to cope with administration and the challenge of aligning the global issues with local struggles.

Conclusion

If transnational social movements are understood as complex social systems of coordination and control that arise and exist within communicative practice, the 'constitution question' is a fundamental aspect of organisational communication (Koschmann, 2011, p. 5). Social movements defy conventional organisational design and knowledge codification. They may appear to function as coherent units of action and intention, masking hidden tensions and contradictions, yet over time they develop ways of working that are amenable to organisational analyses. In the case of LVC, the ideational pathways between the issue-specific campaigns of individual member organisations can be traced back to the self-structuring master frame of food sovereignty, while activity coordination and institutional positioning are governed by local constraints and enriched by local cultural politics. Membership offers a further layer of complexity as individuals experience collective identity with their local organisation and, simultaneously, with the wider LVC movement.

Analysis of communication flows can expose the 'critical and interpretative edges' (McPhee & Zaug, 2000, para 20) of organisational constitution by explicating the implications of communication episodes within organisations, and by providing a template by which to detect, diagnose and assess novel processes. The four-flow theory provides a roadmap for unravelling this complex identity politics. Attention to the variety of message flows and their different audiences can help to explain how members define themselves and identify with the concept of food sovereignty, their individual roles within the organisation and their perceptions of influence and agency within their own local environments. It also helps to articulate how the movement grapples with decision-making and implementation within a non-hierarchical structure. The third and fourth flows enable us to focus on the independent, interdependent, and collaborative work of members in goal-oriented activities that serve their own purposes and, potentially but not assuredly, those of the broader movement.

The CCO perspective reveals LVC not as a given but as a becoming – a diversity of autonomous organisations of farmers, fishers, indigenous and migrant workers' associations acting locally and also communicatively constituting interlocking coalitions that scale-up to national, regional and transnational mobilisation in joint action around a shared agenda. Emerging through local episodes of communication the organisation is constantly '(re)produced, (re)incarnated and (re)embodied in local interactions, and thus subject to change and renewal' (Cooren et al., 2011, p. 20). Technology-enabled agency enables some participants more choice in constructing their individual forms of involvement. These members may act as brokers between local and international networks. They may scale-up activism thorough loosely coupled coalitions or through engagement in more tightly managed campaigns. They often do both, facilitating hybrid, discursive networks situated in broader, dynamic, political contexts that influence the ways their organisations are constituted communicatively.

As discussed in this chapter, food sovereignty is pursued by actors through a range of communicative strategies including the contestation of knowledge, values and rights. The food sovereignty movement itself is a 'process of knowledge creation' (Rumetshofer & Kay, 2016, p. 17). Agroecology, as a central principle of food sovereignty, embraces 'ecology, ethics and equity'. A truly political science, it is not only radical but 'it may deeply shake understandings and feelings about life itself' (Charão-Marques et al., 2017, p. 137). Applying bio-intensive methodologies in workshops in non-conventional food production, for example, the trainers of AgroSano in Oaxaca, Mexico, have a goal beyond teaching healthy growing and eating practices – 'training has become a site to build a material consciousness. If people know how to produce food, they will become empowered' (Curiel, 2017, p. 148). Generating new relationships through food practices, rural production societies engage in growing, seed exchange, recycling, food fairs and alternative energy projects in the transition to better living. Most deliberate and emblematic of food sovereignty knowledge strategies to develop and spread the resistive epistemology of agroecology are the peasant universities, institutes, and schools discussed in Chapter 5. These alternatives to a mainstream education that reproduces neo-liberal ideology and suppresses voices from the grassroots are enabling social movements such as LVC to strengthen their autonomous knowledge systems.

5 Learning as resistance

Models of education can be pathways to mobilisation against the dominant, market-driven paradigm of industrial agriculture. Embodying an 'ecological and social ethics' (Altieri & Nicholls, 2017, p. 236) the agroecological approach to farming elaborated in this chapter presents methods that counter the top-down modes of operation witnessed in conventional, agricultural extension programmes that support the corporate food system. These methods are taught in Institutes for Agroecology in Latin America (IALA). These schools connect theory to practice, embedding problem-solving in the rural communities in which students live. They link student investigations of local challenges to broader social change movements such as the struggle for food sovereignty, and thus make learning a political act that interrogates issues of power, scale and history, recognises different cosmologies and celebrates rather than elides difference.

The ongoing tension between the technically focused agricultural extension systems and the more socially activating approaches is just one element to consider when examining the role of learning in shifts in agricultural values and practices. A complex cultural politics underpins the shift to agroecological farming, which faces barriers including historic land use and family resistance (Meek, 2016). These tensions remind us that it is important to avoid conflating the individual with the collective in our analysis of any movement (Wolford, 2010). Just as 'no shared cosmology or shared political program' (Grey & Patel, 2014) can accommodate all interpretations of food sovereignty, a single curriculum cannot satisfy the unique needs of those seeking to challenge modernist notions of power and autonomy in the food system.

This chapter applies a political economy lens to the grassroots communication strategies of three La Vía Campesina (LVC) member organisations: the Brazilian Landless Workers Movement (MST), the National Association of Indigenous and Rural Women (ANAMURI) in Chile and the National Association of Agricultural Producers (ANAP) in Cuba – a country whose turbulent political history has resulted in the development of a unique model of organic agriculture. In these examples resistive pedagogical and communication models are applied with the aim of making farmers and all citizens technically proficient, politicised individuals who can help their communities sustain, or make the shift to, ecological farming.

Framing food sovereignty

Resisting simplistic linkages between population growth, climate change, conflict and resource scarcity, civil society campaigns against biopiracy, land grabbing and genetic modification remind us of the unequal power relations in food systems and the non-neutrality of technological solutions owned by private actors. The NGOs and social movements driving these campaigns provide 'access to information, advice and analysis [and] access to debate on areas that involve political choices' (Murdock & Golding, 2005, pp. 65–66). Defending producer and eater rights to information against the incursion of powerful ideologies such as free trade, and against powerful actors such as agribusiness companies, the movements seek to disrupt and reorient the power relations that underpin production, distribution and marketing of food through ideological contestation. They oppose what David Harvey (1996) calls dispossession by accumulation – in this case the exploitation of small-scale farmers by profit-making corporate actors. They aim to demonstrate that cheap food is not so cheap when its social, cultural and environmental impacts are taken into account (Rundgren, 2016; Patel & Moore, 2017).

Food sovereignty, introduced in Chapter 4, is an accommodating master frame for the mobilisation of LVC as a transnational coalition (Mann, 2014a). The movement's members argue that neo-liberal policies contribute to hunger and poverty, 'prioritis[ing] international trade, and not food for the people':

> They haven't contributed at all to hunger eradication in the world. On the contrary, they have increased the peoples' dependence on agricultural imports, and have strengthened the industrialisation of agriculture, thus jeopardising the genetic, cultural and environmental heritage of our planet, as well as our health. They have forced hundreds of millions of farmers to give up their traditional agricultural practices, to rural exodus or to emigration.
>
> (La Vía Campesina, 2003)

A utopian alternative to neo-liberalism, food sovereignty is problematised by the relational nature of sovereignty. Does it operate on the scale of the nation, region, or local community, or is it a 'peoples' sovereignty' and, if so, what does this mean? (Edelman, 2014; Patel, 2009b). Arguably, popular sovereignty has been adapted or 'abused' by nation-states to accommodate shifting relationships within global markets while corporate actors have become sovereign powers 'akin to powerful feudal lords' in the neo-liberal era (Conversi, 2016, p. 488; see also George, 2015). This scenario is complicated by sub-state nationalism and territorial claims of indigenous peoples, which remain unresolved but essential to the concept of food sovereignty in many nations including Australia and New Zealand (Mayes, 2018). The ontological basis of food sovereignty lies in respect for different knowledges and ways of knowing, therefore it necessitates encounters between groups who have historically been in conflict over territory, such as pastoralists and indigenous people. Further, while the struggle for food sovereignty is rooted in conflicts over land, soil and territory, it is increasingly relevant in cities where urban eaters now contemplate the impact of climate change on food supply (McMichael, 2014). All these elements give rise to complicated, often overlapping, sovereignties.

In LVC's framing, food sovereignty celebrates these complex identities and evokes a transformative, democratic style of politics that offers a stark alternative to the vertically organised industrial food system. It is employed by member organisations to suit specific contexts, assuming national, ethnic, universal and/ or cosmopolitan features (Patel, 2009c). Somewhat duplicitously, it suggests nations assume control over their food security policies, including the right to impose protective tariffs against the dumping of subsidised exports and the support and promotion of local markets. It also puts the onus on governments to respect, protect and fulfil the rights of citizens to food and the productive resources to produce it, including land, making it part of a larger project to overhaul the neo-liberal economic model.

Fundamental to the political project of food sovereignty is the principle of agroecology, which operates as 'an alternative value system' (Meek, 2014, p. 48) with which to counter the industrial model of agricultural production. Defined as 'a way of redesigning food systems, from the farm to the table, with a goal of achieving ecology, economic, and social sustainability' (Gliessman, 2016, p. 187), it is both an 'objective and a strategy … not only a means of production but also a praxis of change' (La Vía Campesina, 2017b, p. 36). It allows for the building of autonomy from the production systems of the hegemonic model, resists and deconstructs the dependence on commercial seeds, pesticides and fertilisers which are becoming increasingly expensive, and opens up 'the possibility of building and salvaging knowledge which is part of a sustainable relationship between humans and nature, based on ancestral knowledge, culture, and territorial diversity' (ibid.). It is part of a larger movement towards an 'emancipatory rural politics' (Scoones et al., 2017; Andrée et al., 2014). Through practising agroecology, 'the application of ecological concepts and principles to the design and management of sustainable agricultural ecosystems', it is claimed that small-scale farmers can lay the foundation of an 'epistemological, technical and social revolution … from below' (Altieri & Toledo, 2011, p. 587). According to advocates, the approach also represents a solution to declining local economies and rural unemployment. As a response to a resource intensive and environmentally destructive conventional production model that relies on fertilisers, seeds and herbicides as well as high levels of regulation and certification, agroecology is affordable and sustainable. Its methods are knowledge rather than input-intensive, 'improve links between the land and consumption', reduce waste and risk and empower producers (Lang & Heasman, 2004, p. 32).

Further, advocates promote agroecology as a way to mitigate climate change. They argue it returns energy to the soil which helps to 'cool the planet' whereas industrial agriculture – along with corporate food consumption – relies on intercontinental food transport, intensive monoculture production, land and forest destruction and the use of chemical inputs – all of which contribute to global warming (La Vía Campesina, 2015). They are equally critical of Big Food's 'false solutions' to the energy crisis, which include biofuel production, carbon trading schemes, GMO technologies and land grabs. In response to the Eighteenth Conference of Parties (COP 18) and the Eighth Meeting of Parties (CMP 8) of the UN Framework Convention on Climate Change (UNFCCC) in Doha in 2012 LVC warned of 'climate-smart' solutions such as 'sustainable intensification' designed to increase the yield per unit of land:

> This will open the door for carbon markets in agriculture, will allow for carbon accounting to determine agricultural policy, will open the door to the further propagation of genetically modified organisms (GMOs) and other technological fixes such as synthetic biology, and will favour agribusiness over small farmers and peasants.
>
> (La Vía Campesina, 2012)

In Latin America the academic 're-emergence' of agroecology occurred in the late 1970s and early 1980s as anthropology, ethno-ecology, rural sociology, development studies and ecological economics supplemented a field dominated by agronomy and ecology (Altieri & Nicholls, 2017). In the 1990s NGOs concerned with the social and environmental impacts of the Green Revolution promoted agroecology as a participatory approach to the conventional model of agricultural extension which was based on disseminating agricultural research from technical experts in a top-down manner (Altieri, 2002). These NGOs, led by initiatives including the Centre for Education and Technology (CET) in Chile and Centres for Family Agriculture and Agroecology in Brazil trained hundreds of NGO technicians and farmer-promoters. The Latin American Consortium on Agroecology and Sustainable Development (CLADES) and the Latin American Scientific Society of Agroecology (SOCLA) developed research programmes and progressed the agroecology agenda in the 1990s producing many world leaders in the discipline. Today, agroecology is described as a 'cognitive, technological and social' revolution of 'global proportions' which has spawned 'new modes of communication between activism and science', in part through its adoption by LVC (Martinez-Alier, 2011, cited in Altieri & Toledo, 2011, p. 589).

Resistive epistemologies in action

Agroecology is a resistive epistemology that 'resumes the indigenous, black, feminist, anti-colonial and anti-imperialism struggle of more than 500 years' and 'defends the great popular diversity of humanity, biodiversity as the organising principle of Mother Earth and the plurality of knowledge' (ANAMURI member cited in Garrido, 2016). As an ideology opposed to the expansion of capital, oppressing patriarchies and hierarchical food systems, agroecological knowledge must be disseminated in a horizontal and experiential manner, where spaces for the learning process are opened up, for 'if the practice is imposed and didactic, instead of endogenous and participative, it contradicts the democratising potential that this social-economic and ecological approach has ... converting [it] into another form [of] epistemological imperialism' (Chohan, 2017).

In LVC's National Institutes or 'schools' of Agroecology throughout Latin America the education of teachers is not limited to curriculum development but 'speaks from critical perspective to the way knowledge is produced in society and how this process can contribute to either merely reproducing relations of power or to the creation of new knowledge and to the transformation of society' (O'Cádiz et al., 1998, p. 89). This methodology is based on the teachings of Brazilian scholar Paulo Freire, leader of the Movement for Popular Culture in the 1960s and author of *Pedagogy of the Oppressed* (1970). Freire established the 'literacy circles' programme with sugar cane workers in Pernambuco, North Eastern Brazil, teaching workers to read and write, an initiative that spread across the country. In his brief tenure as Education Minister of Brazil, Freire worked with local communities to establish the Popular Public Schools 'built on participative planning and delivery' with support from civil society groups including NGOs and social movements (ibid., p. 209).

Aiming to revolutionise educational practices, Freire sought to eliminate hierarchy in the classroom by bringing students in as equal participants in the learning process and insisting that the curriculum must reflect and respond to the lived experience of learners to enable them to develop 'an awareness of the dialectical relationship of local and global contexts with an orientation towards action' (Bolin, 2017, p. 757). Linking learning to cultural politics and class struggle, Freire encouraged students to challenge the dominant ideology through critical engagement or 'conscientization' (Freire, 1998) – a three stage project involving investigation, thematisation and problematisation of the practical needs and daily concerns of peasant farmers. This serves, ideally, to develop 'understanding of the interrelation of local and global issues' (Bolin, 2017, p. 758). Using this cyclical process to create a common vocabulary and shared understanding, students are encouraged to rethink meaning-making systems and engage in an informed way with issues identified in and by the community. This directly contradicted what Freire termed the top-down, teacher-as-expert, 'banking approach' to education that feeds knowledge down from experts to the rural poor and serves to integrate them into the structure of oppression, rather than 'transform that structure so they can become "beings for themselves" ' (Freire, 1970, p. 47). This critical pedagogy focuses on how education is connected to broader social change, and how schooling itself can serve as an 'ideological state apparatus' (Althusser, 1970). Accordingly, students need tools to reflect on the realities of knowledge production, culture, racialisation and gender identities (Tarlau, 2014) and recognise that the food system is an 'ecological, social and economic system and needs to be viewed as such in all its complexity' (Rundgren, 2016, p. 106).

The Freirean approach helps 'students develop a consciousness of freedom, recognise authoritarian tendencies, empower the imagination, connect knowledge and truth to power and learn to read both the word and the world as part of a broader struggle for agency, justice and democracy' (Giroux, 2010, p. 1). It addresses the need to embrace transdisciplinary strategies that 'incorporate non-academic ways of knowing into knowledge generation activities, acknowledging that certain research problems or objectives require engagement beyond narrowly defined expert knowledge' (Valley et al., 2018, p. 6). This recognition of other standpoints and ways of knowing – particularly indigenous ways of knowing which emphasise the sacredness of food, reflect deep kinship with the Earth and rely on intergenerational transmission of food-related expertise – coupled with the sharing of knowledge horizontally, both locally and in transnational circuits, challenges the 'hierarchical concentration of knowledge production by agri-business' (Cid Aguayo & Latta, 2015, p. 404).

Institutes of Agroecology in Latin America (IALA)

The organisations that comprise LVC recognise the importance of creating spaces and 'learning networks' (McMichael, 2008) that focus on the pedagogical aspects of organising critical to their emergence and maintenance (Tarlau, 2014). Their tools include the IALA as 'agroecological training spaces that democratise debate, knowledge, science and technology ... where training is oriented towards critical thinking and, at the same time, seeks to equip young people with practical tools to build food sovereignty' (Garrido, 2016). The schools were originally established by the Latin American Coordination of Rural Organisations (CLOC) in reaction to dominant systems of education in contemporary universities that produce graduates with technical skills that serve the needs of the corporate food regime. According to Francisca Rodriguez of ANAMURI:

> Unfortunately, the professionals who are being trained today in all the universities are professionals for the system, including the agronomy engineers ... in the past an agronomist was more complete, his practice was together with the peasantry. Today, their practices revolve around large corporations and they are all thinking about how they are going to make money without having any consideration or recognition of the value of the Earth, the knowledge of the men and women peasants, the peasant practices for doing agriculture that are amenable to the earth and the environment ... we say no to the technological practices of today that are destructive and devastating towards the earth and the ecosystems.
>
> (Cited in Paget-Clarke, 2014)

The first school was set up in Brazil in 2005 followed by Venezuela (2006), Paraguay (2008), the Amazon (2009), Chile (2014) and Colombia (2016). Argentina, Nicaragua and Ecuador have also established one, and a new proposal for a Haiti school has been put forward (Chohan, 2017). In the schools, agroecological knowledge is grounded in social and environmental justice concerns to create a new ethics where peasant praxis and collective action are valorised. Staff are politically active, and their own organisation, collective bargaining and representation encouraged. Teaching explicitly promotes 'epistemological resistance against agribusiness dominance' (ibid.). While the schools are united in their aims to integrate youth into an alternative food systems paradigm that supports small-scale peasant farming and collective agriculture practices, differences are based on the needs specific to each country and region.

Brazil

An emerging economy with unrivalled rates of inequality, Brazil has one of the highest concentrations of property ownership in the world, stemming from colonial land grants, historical laws of tenure and exacerbated by agro-industrial policies and development projects. The wealthiest 10 per cent of the population possesses 46 per cent of the nation's income while 50 per cent collectively own 13 per cent. One per cent of landowners control 45 per cent of farmland, a legacy of the large land grants to privileged Portuguese families and the institutions of slavery in the colonial era (Carter, 2015). Those who live in the countryside are the most affected by the incursion of international capital.

The Landless Workers' Movement (MST), emerged in the 1980s via religious networks, rural trade unions and civil society activism. Peasant farmers, the urban poor and indigenous members continue to resist the occupation of large tracts of lands for monocultures such as the 'republic of soya' (Fernandes, 2015). In these sites the persecution and criminalisation of protestors and human rights defenders is common. According to the Pastoral Land Commission, conflicts over land and environmental disputes lead to 70 deaths in Brazil in 2017 alone, the highest rate since 2003 (Watts, 2018).

In the mid-1990s MST started targeting large global corporations to protest their growing influence in the countryside, recognising the green deserts of monoculture as a new set of obstacles to constitutionally-endorsed land reform – since the 1980s, 350,000 landless families have secured land rights through occupations of unproductive estates (Meek & Tarlau, 2014, p. 246). In the Pontal do Paranapanema region of São Paulo State, for example, hundreds of landless families wait patiently to be settled on a parcel of land. Surrounded by fields of sugar cane, they shelter in roadside encampments, described as 'rural favelas' by an unsympathetic media. The framing of members of MST as fundamentalists and a dangerous menace is a response to the growing role of the movement as a leading critic of neo-liberal policies and its role as a voice for the rural and urban marginalised. Newly elected President Jair Bolsonaro has declared that '"crooks" from Brazil's land workers movement MST would be designated as terrorists' (Phillips, 2018).

In the 1980s, MST observed that the occupation of land would need to be supported by a parallel occupation of the school system to counter these negative framings and the government's increasingly narrow focus on urban priorities. Additionally, educational outcomes across the country reflected 'serious problems' in terms of access, consistency and quality; in the 1990s just under 50 per cent of Brazilians had two years of schooling, considered the minimum for functional literacy (O'Cádiz et al., 1998, p. 47). A thriving private sector funds the education of middle-class children while poor and working-class families endure 'a demoralising public schooling or total exclusion' (cited in ibid., p. 48). Freire's introduction of the Popular Public Schooling or the 'education of the people' transformed the original model of public education to one that engages marginalised groups in critical awareness and action. In rural areas, Education of the Countryside, based on Catholic liberation theology (see Boff, 1994) and Freirean-based study groups, was developed to advance MST's political struggle to obtain agricultural land and promote peasant agriculture as an oppositional territorial paradigm to agribusiness (Meek & Tarlau, 2016; Fernandes, 2015). It was the basis of the Program for Education in Areas of Agrarian Reform (PRONERA) in 1998 and a series of national guidelines in 2001 and 2008 before an Education of the Countryside office in the Ministry of Education in 2005 was created along with a presidential decree of support in 2010.

The Federal Institute of Pará-Rural Campus of Marabá (IFPA-CRMB) was established on an MST settlement in 2007, with the objective of educating a 'critical citizenry that is capable of understanding the social, economic, and political contexts of their home community and its relations to the state' (cited in Meek & Tarlau, 2016, p. 249). A vocational high school delivering agroecological extension training to settlement residents, IFPA-CRMB encourages cooperative initiatives, worker organising and solidarity economy approaches alongside agroecological land management practices. The curriculum is based on an 'alternating pedagogy' where students share their time between the campus and their own community, conducting place-based research to identify the sources of oppression, violence and dispossession that pre-date settlements. Interviewing farmers about their land management decisions, they engage in experiential learning that legitimises ordinary knowledges and lived experience while also connecting local struggles to the wider food sovereignty movement.

David Meek (2016) describes a final class project 'Improving My Land' which engages with the history of production on the landscape, and especially family histories of land management, with the aim of redressing the mechanisms that have led to the dominance of cattle ranching over alternatives such as diversified crops. The project requires students to consider the social, political and economic impacts of development such as a road, and the existing technical practices of the family. A particular focus is on the rationale for changes in practice:

> How have they maintained their life within that space? For example if your land was already cattle pasture, and ... you got cattle because that made sense. With this understanding of the history of your landscape, and your family's role in its transformation, you'll be able to better describe your vision for that landscape.
>
> (Cited in Meek, 2016, p. 286)

Another example is a seed saving bank project in response to concerns regarding transgenic seeds supplied by corporates in packages of inputs, including pesticide, herbicide and fertilisers. Identifying a lack of traditional seeds in their communities, students interrogate farmer reliance on agribusiness companies for a resource that has traditionally been the common property of communities. Gathering, planting, preserving and sharing local varieties among the communities, the students engage in transformative pedagogy and practice, articulating that 'seed sovereignty is about resisting the market control of common resources, breaking farmers' reliance upon the agro-industrial system and helping them regain their agency over the production, management, and preservation of traditional seed stocks' (Meek & Tarlau, 2016, p. 252).

Chile

Chile's National Association of Rural and Indigenous Women (ANAMURI) provides an example of the intersection of gendered and indigenous standpoints within LVC. The association emerged in 1998 as a grassroots opponent of transnational agribusiness in a country widely celebrated as the neo-liberal success story of Latin America. Since the transition from dictatorship into democracy in the 1990s, Chilean elites have promoted the agro-export model as 'an unquestionable success' while 'peasant families are being driven out of the countryside because they have no land or because the abuse of intermediaries – especially supermarkets – and the absence of local commercial channels does not allow [farmers] to sell at fair prices' (Garrido, 2016). Employers exploit the poor information base available to peasants and take advantage of the lack of mobility of male and female farm workers who are left vulnerable to pesticide poisoning, unfair employment contracts and child labour. Less than a decade ago 75 per cent of the owners of family farms were men and only 25 per cent women. Collectively, they owned just 9 per cent of Chile's farm and forest land (Sepulveda, 2009).

A longstanding strategy of ANAMURI has been the promotion of female leadership through Sustainable Agriculture Internships that include workshops on nutrition, women's rights and personal development. The schools were originally based on this internship model, whereby two weeks of face-to-face learning would be followed by three weeks in community, in three blocks. This curriculum was modified to better suit the needs of the targeted group, peasant women 20–50 years of age, which include family dynamics, caring responsibilities and work commitments. The challenge for the schools is to continue to adapt training modalities for the specific needs of rural women with diverse racial and cultural heritage.

ANAMURI is the main driver of the network of seed curators or 'healers' (the *curadores de semillas*) who protect and share a wide variety of seeds in communities. The healing communicates a sense of care, an agroecological response to the creation of transgenic terminator seeds. The networking activities of the *curadores* challenge the sense of displacement felt by farmers who depend on the generative abilities of the seeds, offering an alternative epistemological model to the homogenisation of practice and product witnessed in the corporate food regime. The linking of identities between gardens, kitchens and peasant farms fosters 'resurgence, transmitting the enjoined agencies of soil, climate, seeds and peasant producers into place-based experiences of food' (Cid Aguayo & Latta, 2015, p. 402). Activities include the planting of organic gardens, farmers' markets, food-tasting and seed exchange fairs. These patterns of exchange may not always operate outside the formal economy but they 'politicise consumption and reconstruct food exchange as a space of trust, solidarity and proximity, against the grain of the industrial food complex and globalised circuits of capital accumulation' (ibid.), as well as serving as a practical measure to protect biodiversity.

The first Agroecological Institute for Rural Women is situated 180 km south of Santiago in the town of Auguinco, a Mapuche name which means 'the sound of water' (Jarroud, 2014). ANAMURI initiated schools specifically oriented to women who are less inclined to participate in the co-educational Schools. One ANAMURI member considers this a consequence of a system where women are both 'destitute of land' and 'robbed of education' (Rodriguez cited in Paget-Clarke, 2014):

> For the parents it is very important that their sons have a degree, they never think of the women ... because it is the man who is going to get married and must have a profession in order to be able to support his family ... they are not realising that, today, as many women as men support families. The man must support the family – that is the established view in this very strong capitalist and patriarchal system.

Further, the dominant education system serves as a 'mechanism of expulsion' from the land as there is 'no school education that allows us to take pride in our knowledge or the work of our parents' (ibid.). Children go to the city for secondary education where the central message is that 'peasant production is neither important nor of good quality' (ANAMURI member cited in Garrido, 2016).

The Mapuche, along with five other indigenous groups (the Aymara, Quechua and Licarantai in the north; the Coya and Diaguitas from Atacama; and the Kaweskar in the far south), comprise 35 per cent of ANAMURI's membership. An Araucanian-speaking indigenous group from the south of Chile, the Mapuche situate their claims in the wider struggles of their people, as opposed to those of the middle-class urban women's movement into which they never truly integrated. As 'marginal recipients of rights' indigenous women are reluctant to embrace feminism (Radcliffe, 2002). Aspects of the feminist agenda are foreign to their culture, do not fit with their concerns and are sometimes seen as bourgeois. The world view of Mapuche culture, for instance, includes a complementarity with rather than a separation from men (Richards, 2004). Claims for cultural recognition clash with the hegemonic socio-economic and ideological goals that underlie Chile's national development, goals privileged above women's rights, indigenous rights, and social equality. As far as the Chilean state has been concerned, indigenous women are subsumed within other categories – as peasants, women or 'ethnic subjects'. *Indigenas* are seen as particularly 'non-modern and non-national', and are not seen as legitimate claimants of rights. Instead their needs have traditionally been largely incorporated into 'mestizo development and masculinist political models' (Radcliffe, 2002, p. 154). In this context schools play an important role in strengthening indigenous identity as well as countering patriarchy and capitalism.

Cuba

Cuba provides a particularly unique case of agroecology in action as its farmers were forced to engineer a radical shift away for chemical and machinery-based agriculture during the so-called 'Special Period' which followed the collapse of the Soviet bloc in Europe in 1989. The revitalisation of traditional farming techniques and removal of external inputs including chemical fertilisers, herbicides and pesticides resulted in a transformation to a model of organic agriculture that is now an exemplar for other countries. Methods of knowledge transmission including *diálogo de saberes* facilitated the diffusion of practices and innovations across the country in an 'agroecological revolution' based on values of collectivism, solidarity and cooperation (Sosa et al., 2013).

The importance of the development of grassroots social processes to build the capacity of farmers was quickly recognised in Cuba following the collapse of the Soviet bloc and the strengthening of the US trade embargo. The farmers' association, ANAP, adopted farmer-to-farmer methodologies to enhance food production in the Special Period. To convert a system previously reliant upon Green Revolution technologies provided by the USSR it was necessary to disseminate knowledge rapidly and effectively. Substituting imports with ecological inputs relied on social movement dynamics, demonstrating agroecology is a political movement embodying 'economic, social and ethical principles that are required to fulfil the number one patriotic duty of rural farmers, which is to produce for the people' (Lugo Fonte, cited in Sosa et al., 2013, p. 91). Further 'massification' through school and communities as well as campesino networks became a key element of the methodology.

Horizontal training systems in farmer education schools such as the Niceto Perez, founded in the province of Villa Clara in 1996, and Cuba's specific Farmer to Farmer Agroecology Movement (MACAC) drove the diffusion of knowledge, later extended to the University of Havana where seminars in agroecology and sustainable agriculture were taught in person and via correspondence. Leaders of ANAP participated in environmental policy courses alongside farmer-to-farmer methodology to learn about legislation. In 2009 provincial coordinators of all municipalities were taught, and the Niceto Perez has delivered education to other countries in Latin America and the Caribbean.

One of the most noteworthy achievements of ANAP is the extension of the MACAC throughout the provinces, coordinated through the ANAP infrastructure. Diffusion of the movement was accelerated beyond pre-existing farmer networks through newly appointed coordinators or 'cadre' of ANAP who serve as connectors between geographically distant groups and, being skilled at organising working groups, 'streamline' training. A further innovation was the formalisation of the 'Five Steps' of farmer-to-farmer methodology. The first step relies upon participation on the ground including identification of problems, the sharing of experiences between a farmer and a 'promoter' – another farmer who has implemented a solution on his or her farm. This is the beginning of an ongoing dialogue, step two, as one solution might not suit another farmer's circumstances. The third step involves methodological tools such as workshops, exchange training sessions and farm visits to develop solutions appropriate to different contexts. Experimentation and innovation to broaden the spectrum of techniques is the fourth step. The final stage is reflection and feedback, to review achievements and set new priorities. Further refinement of the methodology such as the Banes Method, named after its province of origin, enabled the systematic identification of the problems on specific farms to facilitate more targeted training and problem-solving (Sosa et al., 2013).

Driven by necessity, the growth of the agroecology movement was faster in Cuba than that throughout other Latin American countries. The number of farming families participating in MACAC reached 110,000 in the decade 1999–2009. Between 2002 and 2008, the number of campesino promoters grew from 1,600 to 11,935; facilitators from less than 500 to more than 3,000; and coordinators from 14 to 70 (Sosa et al., 2013, p. 105). While the Special Period provided initial urgency to the transformation of Cuban agricultural systems to agroecology, many supporters joined the movement in years of economic growth, attributed to the 'dynamism … and persuasive power' of the movement. Undoubtedly, the highly organised ANAP network provided the infrastructure, funding, and the ideological foundations for an 'incubator for the mass movement … highly organised at the grassroots, with many members' (ibid., p. 104). It has since been claimed that agroecological plots are more resilient to the impacts of hurricanes such as Mitch in 1998, retaining more fertile soil and vegetation through contours planting, gully control and greater ground cover. Study author Eric Holt-Giménez (2002, p. 93) reported 'agroecological plots consistently have more topsoil, less erosion, more vegetation and lower economic losses than conventional plots' and recommended that 'participatory, sustainable reconstruction should be a high priority for relief and development efforts' (ibid., p. 103). Hurricane Ike in 2008 left industrial monocultures 'generally, entirely destroyed and in a poor state of recovery' while agroecological farms lost just 50 per cent of their productivity and visibly began to recover within 40 days (Sosa et al., 2013, p. 121).

Invited spaces for policy-making

The mitigation of the effects of climate change has become an important argument for supporters of agroecology in global arenas. According to the FAO,

> agroecology is fast gaining interest worldwide among a wide range of actors as a holistic response to the multiple and interrelated challenges facing food systems – not least of which include continued poverty and hunger in the context of degrading natural resources and climate change.
>
> (FAO, 2018a, p. vii)

Recognition of the potential value of non-expert knowledge, and the explicit focus of agroecology on social and economic dimensions of food systems, the right to food and the focus on equity, gender and indigenous issues, has further contributed to an opening up of 'invited spaces' in global policy arenas. As described in Chapter 4, in 2009 the International Planning Committee on Food Sovereignty (IPC) facilitated the participation of LVC and its allies in international governance through the reformed Committee on World Food Security (CFS, 2012; Duncan & Barling, 2012). Supported by reports from the Food and Agriculture Organisation (FAO, 2015a) and independent scientific studies including the *International Assessment of Agricultural Knowledge, Science and Technology for Development* (IAASTD, 2009; see also Valenzuela, 2016) agroecology is now recognised as a possible alternative to a second Green Revolution and as a basis for achieving food sovereignty and protecting farmers' rights, including the rights of crop producers, livestock keepers and pastoralists, fisherfolk and local and indigenous communities to natural resources such as land, water, forest and genetic resources. The 2014 Symposium on Agroecology for Food Security and Nutrition held in Rome and a general meeting of the IPC for food sovereignty in Gujarat, India in August 2015 highlighted that small-scale producers everywhere, and especially those in the Global South, bear the brunt of the negative impacts of trade liberalisation, land grabbing and climate change. Regional meetings on agroecology, such as that held in the Asia Pacific Region in November 2015, have aimed to develop support for agroecology in locations where traditional agricultural methods are threatened by the move to industrialised food production systems. At that particular meeting it was agreed that supporting agroecology is a way of protecting cultural heritage, food sovereignty and local rights to natural resources in the region. Importantly, it was noted that agroecological initiatives can play a significant role in the reduction of rural poverty and the eradication of hunger and malnutrition while enhancing the resilience of agriculture against climate change, promoting sustainable agricultural development and offering employment opportunities for rural youth in island nations. All these factors contribute to the prospering of alternative livelihoods in the food production system and can help prevent the rural exodus taking place in Asia and the Pacific (recent estimates indicate that some 46.5 per cent of the region's population now live in urban areas, up from 23 per cent in 1970) (Hugo, 2014, p. 5). At the meeting it was proposed the FAO set up a new regional initiative on agroecology that includes a monitoring system of all activities of FAO and governments in the region. The FAO launched the Scaling up Agroecology Initiative at its Second International Symposium in April 2018 (FAO, 2018a).

While this wider, formalised engagement with the agroecology concept is welcome, LVC is equally wary of the co-option of the concept by elites including the World Bank who have enabled corporate actors to successfully enter, and dominate, niche organic markets and capture the concept of sustainable development (Martínez-Torres & Rosset, 2014; Mann, 2013). This capitalisation of nature aims to reconcile economic growth with respect for natural limits. Sustainability has become central to the discourse of 'ecological modernisation' that promotes proactive regulation to safeguard both the environment and economic growth (Hajer, 1995). It is a discourse that relies on scientific evidence that reinforces rather than challenges the capitalist economic system and its persuasive 'win-win' rhetoric regarding the efficiencies of ecological control (Wright & Nyberg, 2015). Corporations have adopted sustainable development as a means to 'legitimate a global grab to manage all the world's resources' (Harvey, 1996, p. 382). According to this discourse environmental management becomes reliant on government policy and is therefore subject to the hierarchy of powers; looking after the environment is good for business and 'only minor adjustments to the market system are needed to launch an era of environmentally sound development, hiding the fact that the economic framework itself cannot hope to accommodate environmental considerations without substantial reform' (Escobar, 1995, p. 197).

The social movements are wary that institutions including the FAO may dismiss the 'transformative potential' of agroecology and instead focus on it as 'a way of making industrial agriculture less unsustainable' (Giraldo & Rosset, 2017, p. 2). For LVC agroecology is a 'multidimensional space of social processes, sharing, culture and art' that should be led by farmers and their families. In these processes there is no place for biotechnological solutions as 'agroecology is incompatible with genetic engineering, there can be no agroecology with agrochemicals or with the transnational agribusiness corporations' (La Vía Campesina, 2017b). This wholesale rejection of biotechnology is a strategic means to 'challenge the ideological system that protects the corporate food regime' (Gliessman, 2015, p. 310).

Conclusion

Communication is one of the primary sources of resistance to power, particularly in the context of class, labour and land struggles in Latin America (Mosco, 2008; see also Mattelart, 2000). It has underpinned the movement for food sovereignty, which proposes a transformative economic and political framework that demands culturally appropriate food produced through culturally appropriate agroecological methods based on principles of pluralism and inclusivity. The movement aims to achieve the right to food, and end food violence through dialogue, collaboration and cooperation with those experiencing it. It offers a roadmap in a time of crisis, and a powerful counter to the dominant discourse of corporate elites who propose to 'feed the world' (Schechinger & Cox, 2016). Analysing food sovereignty through a political economy of communication lens reveals how different values about food are produced, distributed, exchanged and consumed. Power within the global food system is revealed to be the product of structures and processes governed by actors who seek to commodify food. Structuration processes that engineer social relations around class, race and gender demand researchers focus on the role of social movements in contesting hegemony (Mosco, 1996). This includes embracing standpoint theory, alternative ontologies and cosmologies, and questioning what counts as evidence in a policy-making arena where intersectional analyses are secondary to economics.

The shared vision of agroecology, plus the systematic dissemination of agroecological farming practices through the farmer-to-farmer methodology, is an increasingly significant 'socially activating tool' in consolidating and mobilising the food sovereignty project (Martínez-Torres & Rosset, 2014, p. 994). It represents what Mosco (1996, pp. 137–138) describes as a critical epistemological framework comprised of 'multiple dynamic interactions', based on an ontology that stresses 'social change, social processes and social relativism'.

Over 40 peasant agroecology training schools now operate in the Americas, Africa, Asia and Europe. In 2014, the International Year of Family Farming, a series of FAO multi-stakeholder meetings in Latin America, Africa and Asia determined that agroecology should become an integral part of sub-national, national and regional agricultural policies and that governments should be encouraged and supported to mainstream agroecological approaches to food production in their research and development programmes. Pilot projects, marketing strategies, educational programmes and monitoring systems are planned to alleviate the move to industrialised food production that continues to threaten local and indigenous communities in the regions. Yet LVC is alert to the potential fetishisation and depoliticisation of the issue by corporates and political elites. Engaging youth in alternative, collective agricultural practices through a critical food systems education is one way in which social movements such as LVC aim to retain agroecology as a politically mobilising concept rather than 'a few more tools for the toolbox of industrial agriculture' (Giraldo & Rosset, 2017, p. 1).

The emergence of social movements such as LVC is a direct response to the colonisation of the food system. Beyond seeking to understand the system, they recognise that 'the point is to change it' (Wasko, 2005, p. 27). Observing the market has 'abolished frontiers and seeks to impose uniformity on the planet', the movements insist 'human rights don't stop at frontiers; we must globalise them' (Bové & Dufour, 2002, p. 190). Ultimately, they aim for their message to resonate beyond rural sectors to increasingly urbanised populations:

> The struggle for the implementation of land reform cannot be cloaked as a peasants' exclusive need or banner, *but as a social solution for the whole of society's problems.* From this perspective, it would become viable if it were inserted as a claim, a platform of struggle for wide

> popular sectors in our countries. We have to propose changes in agriculture, in the land ownership and in rural development processes, *as part of a wide popular project for our peoples, where there would be a new economic, social and political order.*
>
> (La Vía Campesina, cited in Reitan, 2007, p. 175)

Engaging with 'basic moral questions of justice, equity and the public good' (Golding & Murdock, 1991, pp. 18–19), the movement represents the standpoints of those most affected by hunger and those marginalised, to date, in agricultural policy-making. Declaring that 'patriarchy is an enemy of our movement' LVC recognises that 'the lack of tolerance for diversity is part of the process of dispossession of rural youth ... a diverse, non-violent and inclusive countryside is fundamental' (La Vía Campesina, 2017a). Central to these claims is 'the right to participate fully in social life and to help shape the forms it might take in the future' (Murdock, 1999, p. 8). This right is fundamental in the design and governance of food systems in both rural and urban contexts across the Global North and South as discussed in the next chapter, Chapter 6.

6 Participatory by design

How do we create food environments that respect the rights and dignity of *all* actors in the food chain, from paddock to plate? This is particularly challenging in cities where we tend to see only the consumption and waste stages of the food chain. Thinking about foodscapes outside these narrow categories is new to many municipal authorities and it has not been part of their mandate. But it is increasingly becoming so as urban food production is recognised as a powerful tool to address the vulnerability of our cities to economic and ecological crises, and as an alternative to exploitation under the corporate food regime. As a result, food-centred urban design and planning is becoming an important element in the collective drive to transform food systems.

The new institutional frameworks needed to navigate the challenges must engage civil society actors in governance on multi-scalar levels and also address the urban-rural divide. Venezuela's urban *comunas* provide a working example of *corresponsabilidad*, which is defined as 'a means of bridging the formation of popular power and the existence (and gradual redistribution) of established state power' (Schiavoni, 2015, p. 477). In a nation where 90 per cent of the population is urbanised, this model is rebuilding rural-urban relationships through direct marketing channels and also reframing food sovereignty in terms of eaters' relationship to food and the processes of food production, distribution and consumption. Schiavoni (ibid., p. 476) refers to the *comunas* as 'a demonstration of internal sovereignty in that, through joining together and organizing themselves into a comuna, the communal councils and communities that run them are demonstrating their intent to function as a sovereign unit'. As such, *corresponsabilidad* is an example of the 'interactive governance' (Hospes, 2014, p. 125) that can drive the development of new forms of dialogue between state and civil society actors collaborating on food and agriculture policy. I suggest that this a concept every community can learn from.

Reversing our addiction to cheap and 'convenient' food is not easy but reclaiming public space, including alternative places to eat, produce, prepare and share food, is a starting point for dialogue and resistance to Big Food. Conceptually linking concepts such as food justice, food sovereignty and social urbanism, discussed in this chapter, is a way to weave the 'emancipatory, non-relativistic, cosmopolitan ecology of knowledges' (Santos, 2007, p. xiv) required to re-vision food in the city. This chapter presents planning initiatives that are both political and strategic in engaging multiple stakeholders in processes of co-creating diverse foodscapes. Some embrace notions such as the informal city (McGuirk, 2015), others engage with struggles for land justice that transcend rural/urban boundaries. The range of international examples of food policy instruments and practices presented here is by no means exhaustive but aims to highlight the tensions, contestations and possible bridges to transforming urban food environments through voice, participation and re-learning the city.

Eating under authoritarian capitalism

In affluent Western cities supermarket urbanism prevails. It changes 'the social and physical texture of cities, and with it the very nature of urbanity' (Steel, 2013, p. 145). The large chain supermarket is a clear marker of the death of public space, with its endless array of globally-sourced, processed, packaged and scentless products for 'consuming subjects' (Blay-Palmer, 2008; Johnston et al., 2009). Conventional supermarkets, coupled with the convenience of fast food, are the first response of the free market to rapid population growth and gentrification in cities around the world but their disastrous impacts on our well-being are only just being acknowledged. The US, with rising rates of obesity and malnutrition, particularly among marginalised populations, is a case in point. The Fordist technologies of the industrial food system enabled Americans to turn their 'amazing ethnic inheritance into a fearful, neutered, deeply unhealthy national gastronomy' (Steel, 2013, p. 246) within a century. While supermarkets are not necessarily always the 'bad guys' they are the 'big guys' (Knox, 2015), and they carry the lines of the largest food conglomerates in the world including agribusiness Cargill, who are proud to proclaim:

> We are the flour in your bread, the wheat in your noodles, the salt on your fries. We are the corn in your tortillas, the chocolate in your dessert, the sweetner in your soft drink. We are the oil in your salad dressing and the beef, pork or chicken you eat for dinner. We are the cotton in your clothing, the backing on your carpet and the fertilizer in your field.
>
> (Cargill corporate brochure, cited in Kneen, 2011)

In *Hungry City* Carolyn Steel declares that 'the modern food industry has provided us with what our ancestors always craved: cheap and plentiful food' (Steel, 2013, p. 96). Then she counts the price, making the apt comparison that we are as dependent on food conglomerates, like Cargill, as our ancestors were on kings and emperors. Unlike our predecessors, however, many of us have no direct relationship with those who feed us. This is particularly true of those of us who live in cities. We are hostage to supermarkets offering us the illusion of choice – a range of homogenous products, the freshest of which are artificially ripened, frequently tasteless and devoid of the nutritional value their counterparts provided before soils were depleted by intensive production and monocultures. In the name of efficiency we are offered an ever-shrinking range of produce while being persuaded to accept faux connections between producer and consumer through advertising and marketing campaigns in which:

> Origin, quality, authenticity, freshness and specificity of products, and of associated ways of producing, processing and marketing, are clearly articulated in order to attract consumers and to communicate the distinction embodied in food – the distinction that 'passes' to the consumers themselves (and to the act of consumption).
>
> (van der Ploeg, 2008, p. 279)

We have been conditioned to demand perfect produce – in and out of season – that meets the colour, shape and size preferences of the major retailers. This results in an ever-shrinking genetic resource pool as farmers focus on meeting supplier demand and promotes distancing within the food chain which conceals potentially unsafe, unethical and environmentally damaging production practices despite the co-opted discourses of 'organic', 'sustainable', 'fair trade' and 'alternative'.

Supermarkets themselves are 'non-spaces' (Augé, 1995) that say volumes about our social and cultural relationship with food. Empowerment is reduced to purchasing power and choice is an illusion when shelves abound with products produced by a limited number of 'category captains' who create strong stables of brands and dominate entire ranges of products (Hendrickson et al., 2008, p. 28). Supermarkets have developed 'integrated supply chains, from farm-to-fork, under the prescriptive orchestration of the retailers' (Kjaernes et al., 2007, p. 133), a global trend which results in counter-seasonal produce and a transformation in the range and variety of food available. *The Agri-food Atlas: Facts and Figures about the Corporations that Control What we Eat* (Chemnitz et al., 2017) reports that the food sector has changed dramatically over the past three decades with a shift towards finance capital and biotechnological solutions, as discussed in Chapter 2. Many long-standing players in the processing sector including Cargill (grain), John Deer (farm machinery), Unilever (processed food), Nestlé, McDonald's, Coca-Cola, and Pepsi-co have been joined by competitors from the emerging economies. BRF (formerly Brazil Foods) and JBS, also Brazilian, have acquired three of the biggest meat producers in the US – Swift, Pilgrim's Pride and a portion of Smithfield Foods. The China National Cereals, Oils and Foodstuffs Corporation (COFCO) now owns commodity traders Nobel (Singapore) and Nidera (the Netherlands). The biggest player of all is Brazil's 3G Capital which acquired Burger King in 2010, Heinz in 2013, and Kraft (through Heinz) in 2015. Kraft Heinz is now the world's fifth largest food and beverage company with revenues of US$6.6 billion in 2016.

Despite their influence these giants have paid little attention to their impact on human and environmental health. Their key motivation in consolidating is cutting costs and increasing profits through diversification and venture capital investment to overtake smaller brands. In these relationships 'it's a long way from field to plate … [and] farmers are the most vulnerable link in the chain' (Chemnitz et al., 2017, p. 10). Patterns of investment are rooted in colonial and plantation agriculture, which is mimicked today in the grabbing of land to grow monocultures to feed an industrialised agricultural complex which now includes growing quotas for biofuels. As Chemnitz et al. say: 'Plantation corporations are modern, financially strong actors that are transforming agriculture into agro-industry. We can no longer see them as mere relics of colonialism' (ibid., p. 16).

While farmers are the most vulnerable link in the chain, urban consumers are the most naïve. A large reason for this is that the influence of these corporations can be challenging to detect considering they often sell their products under a variety of brand names through food retailers who are also consolidating. Concentration of ownership increases the power of a handful of supermarkets to set the prices paid to food producers and limits competition from smaller retailers, in turn reducing consumer options. The dominant players also dictate the rules of the playing field, at least in Australia. *The Competition and Consumer (Industry Codes – Food and Grocery) Regulation* (2015) (the Grocery Code) is an industry-led initiative that was jointly developed by major retailers Coles, Woolworths and the Australian Food and Grocery Council (a supplier representative organisation). Woolworths, Coles, ALDI, and Metcash Ltd comprise 80 per cent of the grocery market in Australia, with the two major supermarkets, Woolworth and Coles ('Colesworth') encompassing 65 per cent along with extensive holdings in liquor stores, hardware outlets and poker machines (Rooke, 2018). The negative impacts of this concentration of ownership were noted in the Independent Review of the Grocery Code of Conduct (Samuels, 2018) which reports:

> There has been a considerable effort from both Coles and Woolworths to substantially reduce prices and promote everyday low prices, such as Coles' 'Down Down' or Woolworths' 'Prices Dropped', largely in response to the continued growth of ALDI as well as threat of new market entrants that are expected to aggressively price discount. This has placed significant pressure on small, independent retailers.
>
> (Samuels, 2018, p. 10)

The Grocery Code represents a call for change in business culture in the industry. It was created 'in response to community concerns that the major retailers were mistreating suppliers' (ibid., p. 16; see also Knox, 2015). During the review suppliers attested to 'instances of unfair tactics and arbitrary decision-making with little regard for the potential damage to the supplier's business' (Samuels, 2018, p. 17). Examples include Coles' attempt to claw back $AUD12 million in unapproved rebates from 200 suppliers to cover the costs of 'improvements to supply chains' in 2011 (Samuels, 2018). Woolworths has responded to accusations of bullying with a 'Voice of Supplier' survey while Coles has launched a 'Nurture Fund' to 'support the growth of innovative small suppliers' and has promoted an independent arbitration mechanism. Only in time will it become clear whether these measures are sincere or merely performative.

Despite strong public and media support of Australian farmers struggling for a fair share of profit, purchasing patterns suggest people prefer supermarket produce *because this is what most people eat*. That is a direct consequence of the dearth of alternative foodways available to the average eater in Australia's highly concentrated grocery market. In rapidly growing cities such as Sydney, urban developers rely on the free market to provide grocery options for residents in new housing estates. 'If we build it, Colesworth will come' is a common refrain. Ironically, the major retailers frame themselves as the solution to food waste and food insecurity by partnering with food relief organisations in a solidarity that reproduces inequality and exclusion. The 'corporatisation of food poverty' (Riches, 2018) is highly problematic. While emergency food relief is incredibly important, it reflects a deficit lens that disempowers eaters and also

> relieves the state of its responsibility to provide environmental protection and a social safety net, a responsibility that is particularly important in light of the state's role in pursuing or permitting the destruction of communities and environments in the first place.
>
> (Alkon & Guthman, 2017, p. 14; Valencia, 1997)

In conditions of austerity where the rolling back of social services, the proliferation of non-binding forms of regulation and the devolution of responsibility to individuals for their food choices widens gaps in the provisioning of food, civil society mobilises. Foodbank Australia, for example, currently feeds 710,000 people per month, a quarter of whom are children. Forty per cent of the food and groceries they procure goes to regional areas, including communities devastated by the most severe drought in recent memory (Sullivan, 2018). Despite its benefits, charitable redistribution of surplus food undermines attempts to constrain food waste and also perpetuates stigmas and reduces agency of those most insecure (Campbell et al., 2017; Cloke et al., 2017; Kenny & Sage, 2018). Stigmatisation includes myths that low-income groups lack the skills to grow and cook healthy food; as Dixon (2014, p. 178) notes, 'knowledge of how to eat is not always sufficient for choosing to eat differently than we do'.

Food distribution has become a 'symbolic battlefield' (Chatzidakis, 2014, cited in Calvário & Kallis, 2016, p. 12) in Greece where household food insecurity escalated dramatically under no less than 14 austerity packages between 2010 and 2017. Low household incomes and 'greed and unethical pricing practices' (Skordili, 2013, cited in Calvário & Kallis, 2016, pp. 11–12) on the part of corporates drove the emergence of a grassroots solidarity movement of 'no-middlemen' groups. In 2012 farmers distributed free potatoes in a protest again low prices and retailer tactics of rejecting produce to control prices. By 2014 the deliveries extended to flour, olive oil, legumes, honey and cheese. Over 5000 tons of food was distributed from 2012–14 (Calvário & Kallis, 2016, p. 7). Self-organised groups with open assemblies, making decisions by consensus, included activists, farmers and volunteers. There is no central coordination; each groups operates locally with some examples of regional networks and logistical support through 'Solidarity for All', a structure personally supported by several sympathetic politicians. Distributions, from parking lots and other public spaces, typically offer 20 to 40 products including fresh produce. Consumers pay farmers directly, completely bypassing the retailers.

Issues of space, place and scale are emerging as the distribution network grows. But it is viewed by many as part of a re-localisation strategy for not only the Greek economy, but also territorial governance. The no-middlemen groups claim control of the food system is 'central to the strategy of toppling the neo-liberal austerity agenda', on the basis that 'we cannot achieve political uprising in a country that does not produce enough food' (ibid., p. 13). Are these the seeds of the food revolution we need?

A wonderful revolution

Another example of re-localising the food economy is found in Bologna, Italy. Campi Aperti ('Open Fields') is part of the Genuino Clandestino network which promotes food sovereignty and participatory governance of local food systems. Through participation in this network farmers have gone from price takers (those who have to take the price the market offers) to price 'co-maker'. The 84 producer members set prices at regular assemblies in collaboration with consumers. Operations are based on trust; farmers are not required to sign up to organic certification, which is expensive and open to corruption. Proximity creates reliability of product but Campi Aperti 'extends the sense of community from the physically located (bound to the land) to the socio-economically defined (thus unbound), generating trust throughout trading networks that extend into urban spaces' (De Angelis, 2017, p. 297). Monitoring and filtering membership within the network, producers and consumers interview applicants, visit premises and insist traders bring only their own produce to market. Eaters are welcome to visit farms and are directly involved in regular assemblies where strategic decisions are made including questions of market expansion, economic and political analysis, and opportunities for intervention. As an umbrella organisation Genuino Clandestino 'politicises the practice of participatory guarantees and promotes campaigns' to prevent the sale of state land, for example, in favour of a policy to 'incentive social cooperatives with missions analogous to that maintained by the network' (ibid., p. 299). Scale here is defined by nodes that enter into partnership with one another. De Angelis counts 168–252 hectares used by Campi Aperti farmers, comparable to the average US farm of 180 hectares which is usually under industrial agriculture including pesticides and chemical fertilisers. Outside one of the several markets operating in the scheme a sign reads:

> A revolution is happening: always more citizens look for a healthy, local, non-alienated food, respecting people and the environment. Always new peasants look for a way out of the global markets and organise themselves to sell their products directly in the city. An encounter that can produce social cooperation, democracy and the invention of more socially just modes of production, distribution and consumption of food. A wonderful revolution. Food and land as common goods.
>
> (cited in De Angelis, 2017, p. 301)

Inspired by the food sovereignty movement, Campi Aperti, Genuino Clandestino and comparative initiatives throughout Europe in regions including the Basque Country and Greece (Calvário & Kallis, 2016) are representative of the struggle for autonomy of peasant producers in Europe where the Common Agricultural Policy (CAP) seeks to align agriculture with the global market (Mann, 2013). In creating a 'self-controlled and self-managed resource base' that enables a form of 'co-production of man and living nature that interacts with the market' these farmers reduce their dependency on the corporations that seek to control them (van der Ploeg, 2008, p. 23). Countering the 'accumulation of capital with an accumulation of the commons', their approach recognises the significance of public spaces where people share, communicate, exchange and live together, liberated from the 'character of exclusion from the rights of both use and decision-making, instituting instead schema of open, shared use and democratic governance' (Hardt & Negri, 2017, p. 100).

In re-invoking the commons strategies and interventions like these encourage us to 'read for difference' (Gibson-Graham, 1996). They demand the development of 'alternative standpoints' that 'acknowledge attempts to imagine and enact a food politics that achieves different socio-environmental justice outcomes to those of conventional food systems'

(Harris, 2009, p. 60). This new metrics centres on values like healthy environments and bodies, personal well-being, voice and agency and a sense of community. The pursuit of purpose alongside profit can be seen in a growing movement of solidarity economies (SE) around the world. Defined as 'a socio-economic order and new way of life that deliberately chooses serving the needs of people and ecological sustainability as the goal of economic activity rather than maximization of profits under the unfettered rule of the market' (Quinnones, 2008, cited in Loh & Agyeman, 2017, p. 261), SEs offer a new logic of exchange and governance. As such, these economies – variously labelled new economies, local living economies, generative economies, and sharing economies – are the new growth sprouting from the cracks in capitalism. They are most common in Latin America, Canada and Europe. In the Latin American case they have emerged from social movements under leftist regimes, with notable cases being the National Secretariat of Solidarity Economy in Brazil and the formal inclusion of the SE concept in Ecuador. They span the private, public and third (social) economy and include not only cooperatives but community land trusts, community banks and mutual associations. Cornerstones of SE include a view of human beings as interdependent and essentially social. Solidarity is the core value of social relations, and includes the non-human. Democratic practice is essential to developing the empowerment and agency of community members to influence the conditions they live under, while plurality and diversity characterise the market which encompasses 'many existing forms of exchange, production and ownership that function within different value frames than those of the capitalist market' (Loh & Agyeman, 2017, p. 262).

Heterogeneous and entirely contingent on context, SE practices are shared through trans-local learning between cities cross-continentally and globally. *Translating* concepts and models to local conditions is the essence of trans-local learning as 'lessons can rarely be transferred directly' (De Haan & Maxwell, 1998 cited in MacFarlane, 2011b, p. 146) particularly when individualistic cultures resist the application of more collectivist modes of governance. Food is critical in this learning. It is not an exaggeration to say food is a city's heart. It can be used to feed existing power structures, but it can also be used to feed alternative ones and transform not just the places we live, but how we live and how we see our lives.

Growing food – and justice – in the city

Sydney's inner-city foodscape is home to Two Good, whose 'you eat one, we treat one' principle supplies women in domestic violence shelters with delicious and nutritious meals. For each meal catered for one of the organisation's growing number of corporate clients an equivalent meal is delivered to a survivor – a culture of sharing, as opposed to giving, that embodies the ethics of care missing in the corporatised food system. Using fresh produce from peri-urban growers including Cana Farm, which provides employment for people transitioning from incarceration to the open workforce, Two Good's 'Work Work' team of female chefs prepares delicacies based on recipes from well-known Australian culinary ambassadors including Neil Perry, Matt Moran, Kylie Kwong and Ben Shewry. The Work Work team includes women who are long-term unemployed or living in a refuge. They receive training and post-programme support through a network of supporting businesses. CEO Rob Caslick began Two Good with a single barbeque on which he and a friend prepared a decent, 'clean as possible' organic meal for an underserved community in Darlinghurst on a weekly basis:

> After a while we realised that by providing really good food, we were providing more than just food: dignity, self-worth and equality are all achieved when you offer people a meal that you would eat yourself. We share a meal, not serve a meal.
>
> (Caslick, 2018)

Two Good is an example of what Miriam Williams (2017) refers to as '*actually existing* justice in the city'. She argues for a 'politics of the possible' (Swyngedouw & Kaika, 2003 cited in Williams, 2017, p. 2219) that sits alongside utopian thinking as a way to contend with potentially totalising discourses such as neo-liberalism. Following Iris Marion Young (1990) and Nancy Fraser (1999), Williams focuses on how justice can be understood as resistance to oppression and domination through respect of difference. Everyday forms of growing and creating are noted in her case study of Alfalfa House Organic Food Cooperative, Enmore Road, Sydney, a non-profit business that focuses on not only selling ethically-produced and packaged foods and providing workshops on topics such as zero waste but to 'stimulate community development, foster community spirit and promote sustainable living' (Alfalfa House, 2015, cited in Williams, 2017, p. 2225).

Both Two Good and Alfalfa House are examples of sites of learning within the city as a 'pluralised, multiple learning machine' (McFarlane, 2011a, p. 363). Each enterprise uses food as a vector for 'doing/being/thinking urban life' (Williams, 2017, p. 2226) while 'constituting the just city and contributing to healing and repairing our wounds' (ibid., p. 2228). Further, they are actively building strong relationships with farmers and suppliers, many of whom are within the limits of the city and thus qualify as urban agriculture.

Linking community food security to food sovereignty through social enterprise and urban agriculture lends shape and substance to struggles to define and realise food justice (Heynen et al., 2012). Yet perceptions of urban agriculture (UA) can be a significant barrier hindering its development (Figueroa, 2015; Colasanti et al., 2013). UA practices can be perceived as rural and/or unaesthetic and therefore regressive (Specht et al.,

2016) and can evoke recollections of plantation agriculture and slavery (Figueroa, 2015; Colasanti et al., 2013). Legitimacy can be a challenge when local government officials consider UA as serving social purposes rather than supplying viable quantities of food (Sanyé-Mengual et al., 2016, p. 108). Other barriers to UA-friendly zoning and ordinances are visions of entrepreneurial, global, 'smart' cities (Voigt, 2011; Pothukuchi, 2017; Harvey, 1989; Madureira & Baeten, 2016; White, 2015; Vanolo, 2014). Often depicted as 'pop-up' rather than essential infrastructure gardens are frequently nomadic (Horst et al., 2017; Pothukuchi, 2017; Demailly & Darly, 2017). Visions of UA that strike a balance between urban life, agriculture and ecology are particularly hard to sell in cities biased towards the maximisation of real estate values, such as Sydney, Australia, and Boston, in the US (Loh & Agyeman, 2017).

Integrative approaches to blending policy and planning in the form of enabling ordinances, production regulation and fiscal policy instruments encourage UA activity as well as the creation of structures and processes that facilitate land tenure for production-oriented spaces such as land trusts, preferential tax schemes for agricultural use and developing funding schemes for spaces (Meenar et al., 2017; Horst et al., 2017). Beyond these conventional framings policy can be thought of as a tool that guides the steering of certain social practices at any stage of the food cycle from production to disposal whether this means providing infrastructure and tools, enabling peer-to-peer knowledge exchange and training, or opening up discussions about the potential of UA to build social capital (Spurling et al., 2013). Mitigating financial or bureaucratic constraints, educating residents to forage certain vegetables/herbs from a public communal garden, or setting aside land for community gardens can lead to transformation. Collecting data on how these new production and procurement habits address food insecurity can provide evidence for policy change (Cohen & Ilieva, 2015). Urban green space policy frameworks are becoming more common as municipalities seek to mitigate climate change, provide safe havens for biodiverse species and meet social goals including inclusion through collective food production and consumption. Framing UA in terms of expanding urban green space benefits to include 'recreation and other ecosystem services' is strategic (Contesse et al., 2018, p. 576) and can extend to polycentric governance by public, private and citizen group entities through appropriate policy and procedural mechanisms. However, while the growth in green spaces is welcome, care must be taken to ensure that they are created for and used by the community and not privatised. Further, cities including New York and Paris have witnessed the 'emergence of high-profile "feature" landscape architecture that results in green spaces with high aesthetics but little community value' such as the High Line and Les Halles Park (Erhmann, 2018). The Barcelona Lab for Urban Environmental Justice and Sustainability reports that processes of gentrification displaced vulnerable communities in 60 per cent of neighbourhoods with green spaces built between 1990 and 2008, particularly on the waterfront (Anguelovski et al., 2017).

A comprehensive synthesis (De Cunto et al., 2017, p. 8) of food system policies from more than 49 cities cites identifies community buy-in and local empowerment through active participation in the governance system as key to successful policy, along with shortening food supply chains to make visible 'the socio-economic and environmental relations and connections that shape the urban foodscape'.

Thinking locally, acting *trans*locally

The 'systemic thinking' undertaken in the development of food system policy and programme innovation is, like solidarity economy approaches described previously, shared trans-locally. Localism, according to Sonnino (2017), is inherently trans-local. Expanding conceptions of local foodscapes beyond municipal boundaries has 'important cross-scale repercussions' (ibid., p. 5), including learning and reflexivity in other cities. Convening around good practices and common conceptions of food justice develops capacity to move collectively towards a global sustainable food system. Trans-localism has been adopted as a methodology for 'scaling-out and scaling-up' creative solutions for more sustainable food systems featuring 'citizens' involvement and social innovation, new forms of governance across levels of government, public procurement and collaboration with research' (De Cunto et al., 2017, p. 128). Drivers for citizen-inclusion in urban food strategies include enhancing food security and nutrition; improving livelihoods of urban and peri-urban food producers; employment opportunities and economic development; the protection and restoration of local ecosystems; the reduction of the impacts of a changing climate; and increasing green spaces (De Cunto et al., 2017).

The Milan Urban Food Policy Pact (2015) is a good example of a 'trans-local urban learning forum' that translates ideas and models (McFarlane, 2011b). Milan leads an international exercise in 'Food Diplomacy' that focuses on urban networking, engagement with international agencies, 'seeding food topics' in global forums and project design (BCFN, MUFPP, 2018, p. 22). Food diplomacy is defined as a 'holistic strategy able to increase interests in a more sustainable food system among local authorities and to establish shared spaces of cooperation with different stakeholders committed to this joint effort' (ibid., p. 23). Highlighting the 'innovative, voluntary actions, "flexible tools" capable of providing a place for cooperation, consultation and regional networking among the public and private sectors, together with civil society' the Pact echoes a multi-stakeholder approach that also calls on academic institutions to better understand the complexity of food (in)security and recommends points for strategic intervention through policy. In this context cities are considered says Roberta Sonnino of Cardiff University

> not just as the spatial dynamic through which the interdependent pressures that shape food insecurity converge, but also as places where new food "politics of the possible" are being created and new urban-rural development trajectories can be set in motion.
>
> (Cited in BCFN, MUFPP, 2018, p. 24)

The 179 (at time of writing) cities who have joined the Pact commit to engaging in peer-to peer learning, dissemination of best practice and knowledge exchange and sharing support and advice in developing strategies, charters and partnerships. In this context trans-localism is a site for doing, performing, experimenting, practising and sharing things; for consolidating 'networked relationalities' between food production and consumption (Sonnino, 2017, p. 5). The Pact, emphasising 'community buy-in and system thinking and the preference for a multi-stakeholder approach', facilitates knowledge exchange within 'a local and trans-local network of cities around the world' (De Cunto et al., 2017, p. 89). Reducing food waste, promoting healthy diets and respecting the environment, human rights and worker's dignity were among the 100 concrete proposals for 'local actions that can trigger global change' (Milan Urban Food Policy Pact, 2015; see also FAO, 2018b). These goals – sourced via an intensive social media campaign engaging 1,000 diverse actors (De Cunto et al., 2017, p. 89) – are supported by urban food planners and technology companies following the smart city agenda 'to collect local narratives and to make powerful and socially embedded discourses emerge out of them' (Sharing.Lab, 2017).

In terms of food systems these discourses extend beyond the local to the equitable sharing of the benefits and risks of how food is grown and processed, transported, distributed and consumed; the provision of healthy food for all; ecologically sustainable food systems; fair conditions for food workers; decent profit for producers; and education to create informed eaters. In relating food systems to wider sets of public goods, new spaces of solidarity open up, enabling democratic capacity building, possibilities for more sustainable practices, and an enhanced, cross-scale solidarity.

Re-learning urbanism

'Learning urbanism' is part of 'a project of learning to dwell differently through an inclusive and participatory sharing of knowledge and ideas' (MacFarlane, 2011b, p. 158). The Global North has historically circulated ideas and discourses that promote 'an authoritarian brand of capitalism' or 'neo-liberal urbanism' based on 'anti-poor racial prejudice and crime paranoia' (ibid., p. 143). The 'capitalist imperative to dismantle the old and give birth the new' (Harvey, 2003 cited in McFarlane, 2011b, p. 137) embedded in the discourse of smart, creative cities has produced a form of lean urbanism that is 'fast, technological, research-intensive, aesthetically highly ordered and manicured' (ibid., p. 141). Alternatives to these oppressive paradigms have emerged in 'radical cities' such as those identified in Latin America by Justin McGuirk (2015, p. 23) where incredible urban transformation has been driven by 'strategy rather than style' to cope in conditions of scarcity.

The challenges of growing inequity and social exclusion in rapidly gentrifying world cities where struggles over affordable housing, city land and the urban commons are particularly urgent demand we look to the radical cities for guidance in 'reversing the tide of segregation and rebuilding cohesion' (ibid., p. 27). In terms of integrating foodways into urban planning 'the notion of beginning urban change at the molecular level, through the citizens' digestive tracks' is logical, according to McGuirk (ibid., p. 274). He ponders that health has been neglected by architects since modernism's 'obsession with hygiene' passed, yet notes 'it makes total sense that a community cannot be expected to look after its own interests – economically, socially or urbanistically – if it is dogged by illness and malnourishment' (ibid.). A striking example of food-first philosophy is found at Transborder FarmLab in Tijuana, Mexico, a transnational city with a reputation for violent crime. FarmLab is a food-based project that aims to address malnutrition and unemployment in a 'multidimensional approach to poverty'. It is 'not just about farming and food, it's about ideas and people's capacities' (ibid., p. 275). Generating more electricity than it consumes through solar energy and engaging in 'real recycling', FarmLab is a research site and learning centre grappling with the digital divide 'in a place with no access to technology, in a place that doesn't read and where books are three times more expensive than in the US' (ibid.). Food and literacy are also the focus of library-parks such as Parque Biblioteca Espana in Medellín, Colombia. An integral urban project (PUI), this large-scale renewal of Santo Domingo *comuna* incorporates schools, streets and public spaces co-created with the community. Architects, urbanists, social workers, communications experts, lawyers and a 'social manager' hosted 'imagination workshops' for months, convincing an initially sceptical community of political commitment to the project which has delivered measurable improvements in health, education, employment and well-being.

Outputs of this type of participatory design embody social urbanism, a concept developed by Austrian urban planner Karl Brunner. Designing a city around what is already there, social urbanism counters the modernist impulse to redesign the city from scratch. It is apparent today in planning approaches in Latin American cities where architecture and public spaces are deliberately aligned with social goals in a participative way (McGuirk, 2015). Across the continent nearly a third of city residents live in informal conditions and 85 per cent of housing is built illegally. Or, as McGuirk puts it, 'The favela *is* the city' (ibid., p. 107). In these cities, designing solutions for slums in terms of housing, transport, waste management and public squares is about 'rescuing space as a tool of politics' (ibid., p. 30).

The challenge is not only how to insert services and improve quality of life but integrating the favela into the city, 'creating the connections and flows, the points of communication and inclusion that will dissolve the lines of exclusion and collision' (ibid., p. 26). This requires a smarter and more flexible urbanism that deals with unpredictable changes and includes the entire community.

In Tijuana, immigration focuses municipal attention on the checkpoint between the US and Mexico so communities have taken development into their own hands, turning empty plots of land into affordable housing and social amenities. New housing typologies including small apartments combined with larger houses, live-work units for artists and flexible temporary accommodation are combined with collective kitchens, informal markets and community workshops on a plot called Living Rooms at the Border. A 1920's church, the traditional cultural centre of the community, remains a hub that connects neighbours between plot and town in a dense network of social relations. Housing for the elderly is combined with day care facilities for children in bespoke row houses with granny flats. Diverse land use such as this is almost unthinkable in Western cities where planning is currently done within organisational and disciplinary silos segregating the domestic from the economic, rather than in integrative approaches. The designer of Living Rooms, Teddy Cruz says 'density is still measured as a number of *things* – units – per acre. Why not measure it as a number of social and economic exchanges per acre?' (ibid., p. 279). The Casa Familiar, a non-profit NGO acting as a small-scale developer, serves as 'a conduit between top-down processes and bottom-up impulses'. Enabling the community to become 'co-owner or co-manager of resources' makes sense as 'the paradigm of the top-down is that it just produces units for customers to occupy, but there is no social infrastructural dimension where communities then begin to shape programming, production, food' (ibid., p. 280). While not all the urban design innovations McGuirk focuses on centre on food, the concept of the 'civil imagination' which has been 'sorely eroded' under neo-liberalism is central to redesigning models of governance that can change zoning and enable a 'richer and more empowering community life' (ibid., p. 218). Designing protocols or interfaces between communities and spaces means collective kitchens and markets can co-exist with living spaces. People have access to the tools they need to be economically productive, and have a voice in shaping their community by engaging in 'creative acts of citizenship' (ibid., p. 282). More than deregulation, this is about the top-down facilitating the energy of the bottom-up.

Governance for innovative foodways

In most global cities food and agriculture are incorporated in urban system governance models in hierarchical ways, with varying degrees of participation by residents. Municipal food departments in more developed economies supervise urban food systems and demonstrate a strong commitment to food policy through local food plans and charters based on evaluations like community food assessments (CFAs). CFAs analyse food systems on a variety of scales to better understand how specific UA spaces will integrate into the urban food system (Freedgood et al., 2011). They take into account the four primary avenues for food access for a particular locale – conventional market system; charity systems; federal safety net; and community food systems (Pothukuchi, 2004) – and map the spatial relationships of the food system in relation to areas such as food security, access, distribution, economic impacts, sustainability and urban policy. In the US retail assessments, developed by public health researchers, include the Nutrition Environment Measures Studies which measure the affordability, availability and quality of food in stores and restaurants through a range of constructs such as price, availability of healthy options and promotion of un/healthy food; shelf-space allocation to assess the proportion of healthy to junk food on retail shelves; and Retail Food Environment Indexes which measure access determining the ratio of fast-food/convenience stores to grocery and specialty food stores within a specific area (Minaker et al., 2011). Based on this evidence comprehensive city-food plans can embed UA into long-term planning to achieve specific food system goals and priorities (Horst et al., 2017).

Cities that have developed specific food plans tailored for their given locales include London, Ontario, Canada. Their current 'Urban Agriculture Strategy' (2017) is an ambitious multi-pronged approach to increasing and incorporating UA activity into the city. The Strategy is the culmination of earlier planning documents, including the London Food Plan (2016), the Middlesex Community Food Assessment (2016), Community Garden Strategic Plan (2015), and London Food Charter (2011). It is oriented around nine 'Guiding Principles' that 'complement' rural agriculture; contribute to 'vibrant, healthy, safe and fulfilling' neighbourhoods; prioritise food-based education, training and literacy; act to 'document' local food initiatives; address ecological resilience and biodiversity; provide an 'enabling policy environment' for UA; foster 'community-based decision-making' and collaborative partnership; and support 'diverse communities' in cultivating 'access to healthy, local, culturally appropriate foods' (London Urban Agricultural Strategy, 2017, p. 8).

Cities in Europe are also applying diverse UA typologies that encourage citizens to participate in place-making, which 'capitalises on a local community's assets and potential creating good public spaces that promote people's health, happiness, and economic well-being' (PPS, 2008, p. 5). Paris implemented a permitting process mid-2016 that allows residents to initiate and develop a UA project, such as a rooftop garden, vertical garden or green wall, for a duration of three years. This policy initiative is part of the municipal government's push towards a 2020 target of integrating 100 hectares of greenery into city facades, one-third which is designated as UA (Chow, 2016). The urban greening initiative also involves the introduction of new 'educational farms, orchards and vegetable gardens in schools' (ibid., n.p.). The first set of UA projects involve 32 sites across Paris which are expected to yield, annually, '425 tonnes of fruits and vegetables, 8,000 litres of beer and 45 kilograms of honey' (Reuters, 2017).

The most exciting, and controversial, initiative in Paris is the Agrocité designed by Atelier d'Architecture Autogérée, which offers a vision for the communal regeneration of cities. Containing a community garden, kitchen, conference space and greenhouse, the complex is an example of urban communing. It aims to 'increase urban resilience by introducing ecological transition units managed by the inhabitants themselves within the local environment' (founders Doina Petrescu and Constantin Petcou cited in Drouet & Lacrouts, 2018). The original experiment was evicted from the city of Colombes to make space for a parking lot but now it has found a home in the NW Paris suburb of Gennevilliers. A 'multidisciplinary hub', it offers professional courses about local food networks, beekeeping, composting. Comparable examples of self-sufficient public infrastructure include Madrid, where public spaces function as an urban laboratory for citizen management of energy (de Paz García, 2018). Milan is establishing the 'Open Agri Project' – an integrated hub in the form of an 'urban fringe farm' for training and research, entrepreneurial support and food culture focusing on the entire European Union in a deliberate effort to consolidate the city's international profile as a 'knowledge repository on the city-food nexus' (BCFN, MUFPP, 2018, p. 10). One of the units of the hub, the 'Urban Agriculture Lab', explores UA innovation across 30 hectares of land.

New York's 'Foodworks: A Vision to Improve NYC's Food System' (2010) contains policy recommendations in the areas of production, processing, distribution, consumption, and post-consumption. Since 2011, the city has commissioned a metrics report documenting annual progress in food system goals. Among the highlights of the 2017 report were a strong emphasis on influencing neighbourhood food environments and supply chains to improve the affordability and accessibility of nutritious foods for all residents, regardless of demographic. The amount of school food procurement from local, regional and sustainable sources totalled US$26 million, a 6 per cent increase from the previous year; the number of greater New York City farmers' markets escalated; 27 fresh produce retail outlets were approved in one year (14 completed); and the number of mobile fresh-produce vending carts in low-income neighbourhoods increased to 315. Other highlights include the allocation of US$45 million to provide a resilient energy source for a major food distribution centre that is critical in the case of disaster events; funding for a major food hub to improve local food distribution infrastructure; support for over 540 community gardens and urban farms; and funding for food entrepreneurship. Long-term monitoring of average fruit/vegetable and sugary drink consumption has been instigated in order to assess how food habits are evolving over time in specific demographic groups. These initiatives are combined with a commitment to 'zero waste' to landfills by 2030 which has mobilised organic waste collection and composting by residents and businesses.

A hub for creative growing spaces such as rooftop gardens, facilitated by dense urban infrastructure like Brooklyn Navy Yards, New York City, is also home to multiple projects that tackle serious social justice issues. One such initiative hatched by Drive Change, a social enterprise that empowers young adults coming out of jail, is the Snowday programme. Training formerly incarcerated youths to run a food truck business, Snowday is a highly visible and mobile symbol of opposition to a criminal justice system in which defendants as young as 16 are charged and convicted as adults (the 'Raise the Age' campaign has successfully overturned this legislation – the Correctional Association of New York has announced that by October 2019 the state will no longer automatically prosecute 16- and 17-year-olds as adults). Offering 'savory [sic] and sweet locally-sourced delectables featuring one-of-a-kind New York State Maple Syrup' the Snowday team is engaged in all stages of food preparation, marketing and sales, developing vital transferable skills that will equip them in their fragile transition back to the workforce and society in general. Founder, Jordyn Lexton, thought up the initiative while teaching English to detained teens on Rikers Island. Providing a supportive transition back to society via meaningful work in the food industry struck Jordyn as a way of breaking the cycle of recidivism in New York State, where about 70 per cent of young offenders return to prison within a year of release. The food truck concept also enables the Drive Change team to take on the conventional food system by offering a locally-sourced and ecologically sustainable menu to New Yorkers. Chef Jared Spafford, who brings to the project his ethical food philosophy and broad culinary experience, admits that balancing the ideological principles of sustainable farming and food production with the priorities of the social mission of Drive Change is a constant challenge.

> When a dish sells really well – as in the case of our fried brussels sprouts with apple and sour root – it is logical, from a business point of view, to keep it on the menu. But when brussels sprouts are out of season, the item disappears until the following year.
>
> (Mann, 2014b)

Free, fresh, organic produce from suppliers such as Randall's Island Urban Farm, situated between Harlem and the Bronx, provides opportunities for culinary creativity as the Snowday team devise menu items according to what is available rather than designing the menu and then sourcing the ingredients. Kale pesto, succotash, panzanella salad and cucumber/mint, blueberry/lavender and watermelon/basil slushies are among seasonal accompaniments to popular standards including onion rings, pickled veg, grilled maple-bacon-cheese and pulled-pork sliders. Involving growers in the planning of a menu item leads to further innovation. For example, when the kitchen team envisioned a salad that required tightly bunched lettuce, Randall's Island farm manager Nick Storrs was able to select and grow the type that fitted the bill. Similarly, when Jared wanted to introduce a wider range of pork products to the menu he asked his colleagues at bacon supplier, Flying Pigs Farm, what they found hard to shift. The answer was ribs, which the Snowday team turned into another bestseller using maple syrup. The food truck business is fashionable and highly competitive in New York. A US$7 quinoa salad can be passed over for the $5 plate of nachos offered by the neighbouring vendor. In a city where food is cheap and plentiful (for most), quantity often trumps quality. But Jared maintains that by sticking to the principle of farm-fresh food, grown and cooked well, Snowday is providing an authentic 'farm-to-truck' experience, supporting local New York farmers, and achieving its sustainability objectives. It also returns healthy profits back to Drive Change, fuelling further social justice initiatives.

The logistical challenges of the business, which include satisfying the demands of parking officers and food safety inspectors are typical of the obstacles food entrepreneurs encounter in developing their projects. To reduce some of these barriers cities such as San Francisco have employed top-down strategies. The city has incrementally sought to increase UA since an executive directive issued by the Mayor in 2009 committing all relevant government departments towards improving its urban food system. Highlights of their efforts include the creation of a UA programme that helps develop UA spaces, programming and partnerships; UA resource centres providing garden supplies, educational support and networking opportunities for residents; and revised city ordinances supporting the creation of UA spaces, sales of produce and agricultural equipment (Jones, 2009). As in New York City, the improvement of resident health and nutrition is an important priority. There are efforts to regulate nutrition standards on food and beverages sold in vending machines in terms of fat, sugar and sodium content. Approved in late 2015, the city's 'Good Food Strategy' has overall targets of developing UA so that 30 per cent of produce will be locally grown by 2035 as well as a 30 per cent reduction of food waste by 2020. Their strategy has seven focal points: increasing local sustainable food production and food accessibility; changing food practices of young and elderly people; developing a sustainable and desirable 'good food culture'; reducing food waste; promoting future and innovative food systems; and ensuring strategic implementation. Their urban food system plan focuses on shifting everyday food practices and culture as well as developing technological innovation around data collection through tools like the 'Food Security in San Francisco: Dashboard' created in 2015 to provide a centralised, comprehensive source of data on food-related programmes and activity and baselines on vulnerable populations such as the elderly, children, and low socio-economic groups.

Technology is also a prominent feature of Singapore's efforts to supply a highly urbanised population with fresh, affordable food. Singapore's 'Urban Redevelopment Authority' is considering a range of typologies in its redevelopment planning, including rooftop gardens, vertical farms and floating aquaculture farms. The city's Landscaping for Urban Spaces and High-Rises (LUSH) policy update (of 2017) considers vertical and rooftop greenery – including rooftop farms – as acceptable for 10 per cent of the 'Landscape Replacement Area' requirement. Covered mechanical and electrical equipment that is placed in the top three floors under the roof (as opposed to on the roof) can qualify to be exempt from Gross Floor Area computations as of 2004 if the rooftop features 'activity generating' uses. With the new policy update, rooftop activity can include urban farms and solar panelling. These new policy amendments effectively incentivise and legitimise the use of rooftop farms as approved uses for high rise structures. The previous LUSH scheme resulted in 550 new developments (two-thirds of all residential developments and one-half of 'all malls, hotels, and offices') (Sen, 2017). Elsewhere in Asia, Seoul has dedicated US$46 million to develop urban farming in the city with a goal of 1,800 gardens within a ten-minute walking distance to all homes. Additionally, a special farm zone is to be set up for UA training (Hales et al., 2017).

Finally, Toronto, Canada, has a well-established food strategy that demonstrates how partnerships between diverse actors can build a healthier and more sustainable and inclusive food system. The Toronto Food Policy Council (TFPC), created in 1991 as a sub-committee of the Board of Health to advise the City of Toronto on food policy, is a model of municipal best practice. Contributing to initiatives such as the GrowTO Urban Agricultural Action Plan, Golden Horseshoe Food and Farm Action Plan, Toronto Food Strategy, Toronto Environmental Plan, the Toronto Food Charter and the Toronto Food and Hunger Action Plan, the TFPC plays a key role within the broader national policy framework provided by Food Secure Canada's *Resetting the Table: A People's Food Policy for Canada* (2012). A new project launched in October 2018 and designed to provide food access for Toronto's most vulnerable groups is FoodReach, a community-led, not-for-profit online portal that provides access to quality food at wholesale prices to community agencies and student nutrition programmes that serve over 300,000 low-income individuals and families across Toronto. The platform unites 260 organisations, including suppliers, leveraging the power of group purchasing.

These and many other forms of progressive localism including food business incubators in cities such as Detroit, Michigan and Sydney, Australia (Craven et al., 2018) characterise the policy-making approach of a number of municipal governments faced with a 'policy vacuum left by national policies entrenched within a larger scale productivist paradigm'

(Sonnino, 2017, p. 3). In her survey of 19 documents on food governance innovations in 13 cities across the UK, the US and Canada, Sonnino (2017) identifies four primary values: a view of food as a multifunctional public good; an emphasis on civil society participation and governance; a flexible and inclusive approach to re-localisation; and a focus on the trans-local. Local food plans in these cities share cross-sectoral approaches to food policy that unite climate action plans and transport and health and economic goals, and many emphasise the importance of building a health-focused food system. The most progressive plans also reflect a clear recognition of the impact food consumption has on the environment. Systemic and participatory approaches emphasise the importance of the local scale but 'the re-localisation of the food system is never seen as an end goal; rather it is a means to an end ... part and parcel of wider sustainability strategies' (ibid., p. 4). The critical question, however, remains: what prospects exist for marginal groups to be heard in urban project planning and policy, particularly in contemporary cities where 'conflictual politics and participatory difference are often shunned by a culture of managerialism' (McFarlane, 2011a, p. 368)?

Where participation meets policy

Food Policy Councils (FPCs), such as the Toronto example mentioned above, are typically citizen-initiated groups of representatives working in differing areas of the urban food system that attempt to influence food policy and programming of the city through research, advocacy and education. Across North America, the number of FPCs has risen from 15 in 1999 to 324 in 2016 (Pothukuchi & Kaufman, 1999; Sussman & Bassarab, 2016; Harper et al., 2009). FPCs can play a significant role in brokering policy as a collaborative mechanism that pressures city governments to develop UA-friendly policies and programming (Morgan, 2015) but their influence is limited if left outside of the formal city governance structure and without stable funding. City Planning Agencies are another model acting to integrate urban food concerns within operational areas such as land use, housing and transportation. These agencies can provide important intervention points for engaging planners in food-related policy in overlapping areas such as sustainable communities, 'healthy cities' initiatives and New Urbanism (Clancy, 2004; Raja et al., 2008; Dowling et al., 2014; Morgan, 2015).

City or community food system plans involve gathering key community partners to participate in a visioning and data-gathering process which would culminate in recommendations, benchmarks and implementation steps for improving the local food system in terms of environmental sustainability, nutrition and health, food security, social justice and/or resilience. A 'new food geography derived from urban food strategies' (Duvernoy, 2018, p. 613) is emerging in this liminal space between government and civil society organisations (CSOs). The role of CSOs in promoting marginalised voices and interests in food policy planning and governance is complex, in terms of input and output legitimacy, quality of participation and autonomy, which may be compromised by reliance on funding. But the forms of 'counter-expertise' offered by CSOs can 'enlarge debates and propose alternative policies at the local or metropolitan level' (ibid., p. 616). Coordination and facilitation is vital to building collaboration on various local food system issues. Joint campaigns promoting buying local or eating healthily can be organised to create awareness and momentum towards specific issues. Additionally, planners have worked with youth programmes to create community garden and food entrepreneurship opportunities (Raja et al., 2008).

Stakeholder analysis is vital to understanding the pre-existing, complex, multi-scalar constellation of food system agents, highlighting their values, approach, power source, interests and positions (Campbell, 2004; Houston et al., 2019). This calls for planners to facilitate collaboration between parties with common values as synergy along the spectrum from policymakers to grassroots organisations is needed for engaging with the complexity of the food system. Some cities provide UA resource centres which act as a hub for connecting the various communities, institutions, businesses, non-profits and farmers, and streamline infrastructure support such as facilitating access to utilities (electricity and water through public works), rain barrels and compost. City and state governments can provide dedicated staff support and funding to support institutional 'retrofitting' including the repurposing of public structures such as city halls, schools and libraries. These sites can incorporate UA elements and act as permanent demonstration spaces for the public. Institutional policies surrounding food operations and purchasing practices for governmental food spaces such as canteens, eateries, cafes and vending machines can be revisited to incorporate UA-sourced ingredients. For example, New York City used massive purchasing power of its food procurement systems to change the sourcing of the food ingredients for its public schools to local and regional sources. These changes are also key to the success of Belo Horizonte's Municipal Secretariat of Food and Nutritional Security which helped pilot Brazil's Zero Hunger policies, contributing to decreases in inequality and poverty along with increases in fruit and vegetable consumption (Chappell, 2018).

(Still) a question of land

In our rapidly densifying cities, where space is at a premium, land is key to the foundation for new, inclusive urban ecologies as sites for disrupting conventional foodways. At the city level, land inventories commonly determine the potential of city-owned land to be converted for the use of UA and the most successful operate within existing sustainability policy frameworks (Mendes et al., 2008). Community land trusts are an example of efforts to keep peri-urban farmland viable. Land for peri-urban farming is at high risk in rapidly growing cities. In 2013 in Sydney, Australia, less than 4 per cent of all fresh fruit and vegetable produce supplied to the large Sydney Markets at Flemington was sourced from within 150 kilometres, with 29 per cent coming from regions located more than 1,000 kilometres away (O'Neill & Dimeski, 2013). The Institute for Sustainable Futures predicts that since the current planning system in Sydney tends not to prioritise agriculture as a land use, meaning urban sprawl into peri-urban areas is permitted, the city stands to lose approximately 60 per cent of its total food production by 2031. In other words, 92 per cent of Sydney's current fresh vegetable production could be lost, 91 per cent of meat and 89 per cent of eggs (The Institute for Sustainable Futures, no date).

While the wider food movement has been charged with neglecting the land question UA movements are mobilising around the right to housing, city land and the urban commons. Just as rural land is under new threats from extractive industries, including coal seam gas mining in Australia (Mann, 2018c), 'underserved' neighbourhoods are fighting for access to public land for community gardens, markets and community-building (Kerssen & Brent, 2017, p. 291). Land struggles are linked to climate justice through the issue of GHG emissions in agriculture, the disproportionate impact of climate change on small farmers, land speculation and false solutions to climate change that are used to appropriate community land and resources (ibid., p. 305). These issues are uniting rural and urban dwellers – globally – in novel coalitions referred to as the transnational Blockadia movement. This 'grassroots and broad-based network the likes of which the environmental movement has rarely seen' (Klein, 2014, p. 295) opposes mining of oil sands, the TransCanada's Keystone XL pipeline and the fracking of coal seam gas. Aided by the emergence of a growing sphere of social media activism, local campaigners are creating new opportunities for the scaling-up of concerns and dissent through transnational advocacy networks (Keck & Sikkink, 1998) while participating in local, place-based mobilisations (Cottle & Lester, 2011; Mann, 2014a, 2018b, 2018c). These global discourse coalitions and networks are driven by a common sense of disillusionment with the ability of governments, Big Green and UN Summits to formally address wicked problems such as climate change. In this context defending local landscapes is a way of representing global stakes and demonstrating a desire for deeper democracy. Naomi Klein suggests that

> having the ability to defend one's community's water source from danger seems to a great many people like the very essence of self-determination. What is democracy if it doesn't encompass the capacity to decide, collectively, to protect something that no one can live without?
>
> (Klein, 2014, p. 361; Figueroa & Alkon, 2017)

But the stakes are even higher for indigenous people in settler colonies such as Australia:

> For the First Nations peoples and communities potentially affected by such a vast expansion of mining operations – whose transitional connections to the lands in question stretch far beyond the agricultural concerns of current land owners and related non-Indigenous communities, or the economic interests of government and corporate entities – such projects represent yet another crime in the ongoing history of invasion, forceful land acquisition and resource plundering, and spiritual dispossession, with little to no compensation or benefit for traditional owners, that has comprised the story of colonialism in Australia for the last 227.
>
> (Burch, 2017, pp. 129–130)

Accordingly, land serves as a 'nexus' for justice issues, as a 'potential lever for important policy change', and as 'the foundation for powerful alliances' (Kerssen & Brent, 2017, p. 307) across issues that extend well beyond what we find on our plates. Therefore, as Tanya Kerssen and Zoe Brent observe,

> while it may be tactically safe to limit social justice demands to the products of the food system rather than its means of production, strategically this is a dead end. Farms, gardens, corner stores, markets and food hubs do not exist in cyberspace nor as smartphone apps; they are constructed on the land that is the foundation to the wealth, politics, and socio-economic control of food systems.
>
> (Ibid., p. 308)

Land justice is central to the reconstruction of not only the food system but the public sphere.

Havana, Cuba, provides a stunning example of commitment to UA and associated food infrastructure by powerful political forces. Raúl Castro was an early champion of the movement mandating that the Cuban armed forces participate and grow as much of their own food as possible. Household *parcelas* and *patios, organipónicos* and *huertos intensivos* (cooperatively run organic urban farms) and *centros de autoconsumos* (farms attached to worksites) complement the rural *fincas estatales* (state farms) and private *campesino* farms that provide Cubans with food. The state governs the system under the Department of Urban Agriculture, providing producers and distributors with state support and overseeing the distribution of quotas to local schools, hospitals and nurseries. In 2008 almost 300 cubic km of the 727 in the Havana Province was under cultivation. *Organipónicos* and *huertos intensivos* alone produced over 100,000 tons of vegetables and herbs per year (Gonzalez Novo et al., 2008). While it would be a mistake to characterise the Cuban food system as utopian, the development of UA has made a significant contribution to food security and employment in challenging conditions and presents unique lessons regarding sustainability, resilience and resourcefulness through setting aside land for agroecology in the city.

Conclusion

Recognition of the socio-economic, environmental and cultural inequities of our foodways – and our societies, more generally – is opening spaces for innovation in food planning and policy-making worldwide. In these spaces the notion of food justice is entering into dialogue with the more radical concept of food sovereignty. While both concepts mobilise against the marginalisation of all affected by the 'bads' of the food system, and are focused on less capital-intensive input, fairer supply networks and healthier environments that produce equitable access to quality food, there exist important differences that make the translation of each concept to new contexts challenging. Food justice is a 'contested and complex practice' and 'its strength lies in its ability to engage and bring into dialogue changing discourses and motivations of alternative food systems' (Herman & Goodman, 2018, p. 1042). These systems themselves are sites of 'contradicting values, modern ideas, capital, merchandising, state support and a messy, but not necessarily chaotic, range of agents (humans, other-than humans, and things), with some of them located in other-than-local sites' (Blanco et al., 2017, p. 186). A more expansive and action-oriented concept of food justice (Barnhill & Doggett, 2018; Cadieux & Slocum, 2015) can join many of these dots by bringing rural concerns to the tables of urban eaters.

Any discussion of justice demands a holistic framing and geographic scope that includes issues of power in urban and rural space and production, the common ground of food sovereignty. Agroecology, as a socially activating form of agriculture based on education against oppression, provides a conceptual bridge, particularly when applied in concert with other liberating pedagogies for empowerment. It can support the positioning of food justice as a 'physical and conceptual space of learning, experience and empowerment, in which food can be an effective tool to illuminate and promote greater social equity' (Herman & Goodman, 2018, p. 1043).

Food justice is politicised through the dialogue with food sovereignty. Where the production space is confrontational and demands active agency, knowledge and experience, *campesino* knowledge counts (Milner, 2017). Especially in cities where consumption is at its highest. This acknowledges that 'ultimately food is relational and is always interconnected with its broader environment, which demands that an expansive, responsive and dialogical food justice also accounts for the non-human' (Herman & Goodman, 2018, p. 1044). Therefore, food justice must engage with debates over capitalism, colonialism, climate change, migration, struggles over labour and land, institutional racism, gender and other forms of discrimination. Place-based efforts to transform food systems require 'a strong local character combined with a deep awareness of and participation in global struggles' (González, 2017, p. 155). While local and micro-scale responses are vital ways for people to mobilise and carry out clearly defined actions in response to food-related issues that impact directly on their communities, broader structural injustices to do with land and labour that impinge on the rights and needs of other parties along the food chain must be supported to build collaborations, coalitions and movements. These themes are discussed further in the final chapter, Chapter 7.

7 Conclusion

Voice is never more vital than when 'it seems the utopian imagination is trapped, like capitalism and industrialism and the human population, in a one-way future consisting only of growth' (Le Guin, cited in Tsing, 2015, p. 17). In this book voice is promoted as a value. Following Nick Couldry (2010, p. 2) I 'discriminat[e] *in favour*' of ways of organising systems and frameworks that put valuing voice into practice. This means resisting the forces that seek to undermine and deny voice. These include the institutions and corporations that adhere strictly to the neo-liberal script, evacuating the social from economics, obstructing voice and in doing so preventing justice. I have sought to explain how practices of voice can be sustained and validated in the co-creation of healthy, sustainable and inclusive foodscapes. Meaningful political cooperation that 'unblocks' voices and 'puts voice to work' (ibid., p. 144) is witnessed in countless locally-based examples of collective action, learning, and re-visioning of foodscapes, only a handful of which have been presented in this book.

Giving voice and listening is central to an ethics of care (Cooley, 2010) that recognises the interdependency between the self and the other. The focus on gaining knowledge through knowing others means acknowledging injuries and exclusions. Knowledge is 'best acquired in relation … a powerful source of motivation is an invitation from living others' (Noddings, 1992, p. 113). Building relationships on mutual recognition of voice *invites* participation. The Stop Community Food Centre in Toronto provides an example of the 'strong place-based element' that underpins this participatory ethos:

> Having a physical space that thousands of people can walk into, in which they can sit down for a meal, volunteer, cook, make a telephone call, or connect to community resources is essential … the Stop is a community space where people can have a conversation about food … to generate ideas or contribute. The Stop is continually being approached by other groups that want to reach low-income or diverse communities to 'take their temperature' on everything from poverty-reduction strategies to neighbourhood infrastructure and health. This

> is possible because The Stop is a place where people are already gathering and talking about social issues.
>
> (Scharf et al., 2010, pp. 28–29)

Public spaces such as The Stop act as spaces of narrative exchange that provide opportunities for diverse voices to be expressed and heard. This link between articulation of space and articulation of narrative – what narratives are able to be told in particular contexts, and with what impact – is explored in this final chapter. First, I take up a question already posed by Eric Holt-Giménez – if the food movement is 'rebuilding the public sphere from the ground up' what does this look like? I propose the desire to reclaim voice is common ground for the movements contesting a food system that systematically *disables* voices. I introduce the right to communicate as an essential element of this struggle before concluding with the significant discursive omission in many discussions regarding food justice and food sovereignty in Australia – the dispossession of First Nations people.

Common voice(s)

This book has barely touched on the rich mosaic of food movements mobilising against the corporate food regime through approaches that combine the systemic *and* the local. But one thing is clear; they share many common concerns. These include 'who owns what, does what, gets what, and what do they do with it ... who will transform the food regime, how will it be transformed, in whose interests, and to what purpose' (Holt-Giménez, 2017, p. 232). As self-generated networks grassroots movements carry the potential to unite across common values and collectively generate 'solutions to disentangle what appear to be insoluble dilemmas: poverty, global climate change, terrorism, ecological degradation, polarization of income, loss of culture' (Hawken, 2007, p. 20; see also Klein, 2014). Various forms of 'boundary commoning', operating through communication and cooperation, already connect these issue-based publics. Boundary commoning consists of activities that 'sustain relations among commons thus giving shape to commons at larger scales, pervading social spaces and intensifying the presence of commons within them' (De Angelis, 2017, p. 287). Examples of 'commons ecologies' consisting of 'webs of interrelated commons cooperating at different scales and intensity' (ibid.) are witnessed in many of the examples in this book – movements of peasants, women, indigenous people, communities concerned about the health of their food, their water, their land, their fellow citizens, their neighbours *without* citizenship. All converging against structural injustices imposed by the crisis of neo-liberalism.

An inordinate deal of discussion over the barriers to political alliance occupies us. Movements are made up of human beings so they are naturally fraught with complexity, fragmentation, competition and egotism. Equally, they are infused with goodwill, generosity, commitment and love. Discursive struggles in and between food movements includes debates over whether specific groups are adequately represented; how communications are filtered and centralisation of decision-making avoided; whether movements should unite to create a large-scale united agenda or drive small-scale local coalitions around specific projects (or both); and whether collective action for reform of the food system should also accommodate radical proposals and who should lead these. There is much discussion over the need for actors in the food movement to necessarily possess different strategic orientations. We need 'warriors' to resist the existing system, 'builders' to create alternative approaches to food production and consumption and 'weavers' to create networks and coalitions (Stevenson et al., 2007, cited in Booth & Coveney, 2015, p. 54). These are all important conversations. In fact, 'debates *about* social movement networks largely *constitute* social movement networks themselves' (Juris, 2008, p. 298). Further, activist strategies that aim to be truly transformative and 'supersede social relations of domination through a politics of conflict and social struggle' are necessarily focused on political alliance and convergence on issues, rather than affinity (Calvário & Kallis, 2016, p. 3).

Friction is part of life, and no less part of social movement life (Tsing, 2005; Giraud, 2018). Movements are sites of 'contaminated diversity', of messy, 'encounter-based collaborations' (Tsing, 2015). In this context 'no self-contained individuals or groups assure their self-interests oblivious to the encounter'. This is a vital lesson for those, and I speak here mainly of academics (because I can), who rationalise, systematise and risk colonising movements through their work. As Tsing notes:

> Without algorithms based on self-containment, scholars and policy-makers might have to learn something about the cultural and natural histories at stake. That takes time, and too much time, perhaps, for those who dream of grasping the whole in an equation. But who put them in charge?
>
> (Ibid., pp. 33–34)

This calls for reflexivity to respect and acknowledge the divide between speaking and listening, scholarship and activism and, in doing so, 'counteract the regressive impacts of neo-liberalism' (Sonnino, cited in BCFN, MUFPP, 2018, p. 25) in our own work. We can only be useful to movements if we recognise ourselves as learners united with others in our collective encounters with unpredictable new challenges. In our research we must acknowledge that if 'making voices *matter* is hard; it is even harder, amid the proliferation of new voices, to challenge the hidden forces and dislocations that prevent them mattering when it counts' (Couldry, 2010, p. 50).

It is tempting to imagine a revitalised public sphere can be reduced to a commitment to social justice for the marginalised, an active civil society reflecting that commitment, and an openness to learning through difference. But what is key is the efficacy of distributed voice – how it is recognised and *not misrecognised*. There is no evidence that voice has been heard and had an impact without outputs, or recognition. The status of voice in politics depends on visibility in terms of 'being *regarded as* relevant to the distribution of the speaking opportunities' (ibid., p. 107). The distributed voice, where there is no direct way of linking our inputs to specific outputs, is the essence of a global food politics which operates across multiple scales. It is central to meaningful participation in a food system that includes and respects all voices, involves individuals becoming knowledgeable, sharing and debating ideas, and develops efficacy principles that Neva Hassanein (2008) equates to food democracy.

The right to voice

It is impossible to escape the question of scale in relation to communication given the digital turn and capacity of this technology to connect and empower while simultaneously controlling and surveilling (Lewis, 2018; Mann, 2018b). In the digital era the capacity to form collective, change-engendering narratives on a transnational scale is especially challenging as larger narrative frames subsume smaller stories, stripping them of legitimacy and power. Regardless of the affordances of communications technology available to us it requires a lot of joining of the dots to connect individual stories to wider social and economic injustices – we need sociological and political context to 'grasp the long history of materialisation (the hidden evaluations) that has gone on before someone speaks, or falls silent' (Couldry, 2010, p. 130). Given the explosion of online forms of voice – consider YouTube alone, with 400 hours of content uploaded every minute – storytelling opportunities are plentiful. The limitations of digital storytelling as a form of validation, recognition or narrative exchange in this chaotic, and commercial, space include the incursion of market logics and values inherent in 'self-branding', for profit. Further, it is important to remember that social exclusion is digital exclusion. Not everyone has the financial, social, cultural capital or time to participate. If the creation of a vibrant healthy public sphere is to involve novel communications technologies and platforms these tools must be available to all. Only then can they facilitate the incorporation of different knowledges in concept creation and food systems planning.

To avoid the technologically deterministic assumption that greater Internet connectivity will lead to participation it is useful to consider the history of the emergence of the right to communicate, which precedes the right to food. This right emerged from the concerns of Jean d'Arcy at the UN

Office of Public Information in 1969. A television development leader in France, d'Arcy recognised that new telecommunication technologies had potential to enable ordinary people to communicate horizontally and participate in decision-making in new ways. He predicted that the 'time will come when the Universal Declaration of Human Rights (UDHR) will have to encompass a more extensive right than man's [*sic*] right to information.... This is the right of men [sic] to communicate' (d'Arcy, 1969, cited in Mueller et al., 2007, p. 270). The right to communicate was ahead of its time as a protection against the 'mass media mentality' that supposedly conditioned citizens to accept as 'normal and ineluctable' a top-down flow of information (d'Arcy, 1983, cited in Mueller, 2007, p. 271). Supported by UNESCO and aligned with the New World Information and Communication Order (NWICO) the right to communicate was framed as a collective, sovereignty-based concept. Unsurprisingly it was viewed suspiciously by national governments keen to control news. The MacBride Commission of 1977 endorsed the right to communicate and especially the NWICO needs of countries in the Global South. However its heavy critique of the advertising-based revenue model of the commercial media was rejected by the US under Reagan and in Great Britain under Thatcher. They took the opportunity to withdraw their support from UNESCO and politicised the MacBride report. The right to communicate dropped off the agenda.

Nevertheless, MacBride's vision for the right to communicate incubated in academic and activist circles, culminating in the Boulder Statement (1997) which, sounding very much like the discourse of the food movement, declared:

> The various components of an international movement on media and communications, that can challenge the current neo-liberal orthodoxy, seems to be emerging. The creation of a global social movement [requires] a core constituency of on-the-ground activists who recognise their affinities and can mobilise in concerted actions; an understanding of the key global issues of the day and of the arenas in which they are fought out; and the capacity to get their message out both to natural allies in progressive movements and to the general public.

Fast-forward to 2001 and the UN-driven World Summit on the Information Society (WSIS), created with the express purpose of addressing the digital divide. The Campaign for Communication Rights in the Information Society (the CRIS Campaign) emerged from this summit.

While it has been unsuccessful in its aim of embedding 'the social role of communication at the centre of policy development', the CRIS campaign offers insight into how communication policy can be framed globally and be used to 'mobilise constituencies across borders', according to Mueller et al. (2007, p. 268). As in the case of food sovereignty, the rights frame provides the CRIS Campaign with salience in international arenas where human rights frameworks exist. Both sets of rights are based on ideals of participatory democracy in that they assert that all citizens must be included in every governance process that affects them. In fact the 'right to hear and be heard, to inform and be informed' (MacBride Commission, 1980, cited in Mueller et al., 2007, p. 274) captures the voice claims that are inextricably linked to the right to food. Though, as CRIS Campaign director Séan Ó Siochrú notes, the same tensions remain in that '*even were it enforced* [the right to communicate], it does not address structural inequality embedded with for instance capitalism and its drive to constantly expand its terrain of control' (cited in Mueller et al., 2007, p. 278).

The organisational capacity of the Association for Progressive Communications (APC) connected and coordinated civil society efforts in the WSIS. By providing centrality and a clear hub and spoke structure with the ACP at the centre, and strong horizontal ties between individual actors and advocacy groups, the movement was able to accommodate the shift to later debates on Internet governance (ibid.). This flexibility and scope, coupled with organisational know-how, has similarly sustained LVC as the central and most influential civil society voice in global food politics.

Digital activism is receiving a growing attention by scholars (Schneider et al., 2018; Lupton & Feldman, forthcoming), and has dramatically increased the ability of rooted cosmopolitans across the food movement to transnationalise their strategies (Mann, 2018b, 2014a). Undoubtedly platforms can offer further possibilities for 'progressive forms of learning between different constituencies' (McFarlane, 2011a, p. 360) on local scales. If barriers such as access, education and affordability are overcome, emergent 'networked publics' (Boyd, 2010, p. 39) can provide modes of participation and belonging. 'Imagined collective[s] that emerge as a result of the intersection of people, technology and practice' (ibid.), these publics might function in the manner of offline publics in that they enable people to gather and connect for social, cultural and political purposes, interacting in unique ways that are determined by networked technologies that reorganise information flows. They are 'emergent sociopolitical assemblages with shared or interlocking concerns' (Burgess & Matamoros-Fernández, 2016, p. 81) with the potential to not only engage marginalised groups in the tactical learning that enables them 'to cope with, negotiate and resist' (McFarlane, 2011a, p. 360) the negative elements of the global food system but to exercise voice external of market functioning. However, while technological and social innovation that mobilises 'diffused social resources' including creativity, skills, knowledge and entrepreneurship is vital in the transition to sustainability in our food systems, 'nothing can happen without the direct and creative participation of the people involved' (Manzini, 2011, p. 103). Food-centred practices and activities that occur in the physical communal environment can be 'enhanced but not substituted by online communities' (Vallauri, 2014, p. 173).

Given that the food infrastructure of modern cities has become mostly hidden, contributing to the general apathy of city residents towards the origins of their food (Steel, 2013) an important aspect of information generation is making food infrastructure – from production to disposal – a visible (and multisensory) and vital part of city life again through inclusive workshops, public events and community media development (Dowling et al., 2014). Learning and knowledge exchange can be supported by participatory research methods that are community based and designed to capture and share voice in diverse ways including documentary photography (Alcazar et al., 2017), photovoice (Harper et al., 2017), community coaching (Stulka et al., 2018), and phenomenological approaches (Greer et al., 2016). These methods can help us understand the lived experience of obtaining – and doing without – healthy, nutritious and culturally appropriate food by eliciting and listening to voice.

A different world, from the same window

While my arguments throughout this book are broad in scope and eclectic in theory, I aim to provide sufficient historical and social specificity to make the analysis concrete. To this end it is impossible, as an Australian of European descent, to ignore the paucity of voices of First Nations people in discussions about food in my own country. Here, the subjugated knowledges of First Nations continue to be judged by many as 'non-conceptual ... naïve ... hierarchically inferior' and 'below the required level of erudition or scientificity' (Foucault cited in Mayes, 2018, p. 10). The reactivation of these knowledges, through voice, is essential to 'destabilize and question' the conventional food system in Australia (Mayes, 2018, p. 11).

First Nations scholar Bruce Pascoe, author of *Dark Emu: Black Seeds: Agriculture or Accident* (2014, p. 12) reminds us that 'by adjusting our perspective by only a few degrees we see a vastly different world from the same window'. A growing literature on Australian Aboriginal land practices provides a long overdue case study of the colonising narrative to establish discursive legitimacy for land grabbing. Presented as hunter gatherers to create a rationale for terra nullius, Aboriginal Australians were in fact far from lazy wanderers. They were nomadic by intent – mobility 'let people tend plants and animals in regions impossible for farmers today and manage Australia more sustainably than their dispossessors' (Gammage, 2012, p. 281). While Tim Flannery (2010, p. 100) refers to First Australians as 'ecological bankers', as recently as 2013 *Australian Geographic* magazine managed to write about the arid interior of our country, a 'psychological marker for all Australians', without reference to Aboriginal inhabitation. This despite evidence in the form of explorer accounts that the red centre was well-populated (Pascoe, 2014, p. 33). Flannery draws on Elinor Ostrom's (1990) thesis of Common Pool Resource management to describe how Aboriginal people were 'keystone species of the continent's ecosystems for forty-five thousand years' before 'Europeans displaced them [and] Australia's fragile environment collapsed into a far less productive and diverse state' (ibid.). The totem system whereby each mob has a sacred species to protect, ensured against stress and potential extinction. Deborah Bird-Rose links the balanced, productive and fertile landscape of precolonial times to Aboriginal Australian philosophy of care for the land:

> The state of the country, for instance, offers concrete evidence of the responsibility which the owners have been exercising. Responsibility is grave: there is no hiding in a conscious universe ... the exercise of will in a situation where the choice to deny moral action is to turn one's back on the cosmos and ultimately on one's self.
>
> (Cited in Pascoe, 2014, p. 127)

In discussing the domestication and deliberate cultivation of food plants including indigenous grains, desert raisins or bush tomatoes (*solanum centrale*) in the central desert, Pascoe makes a poignant claim that sums up the colonial attitude to indigenous knowledge: 'The dearth of research on these plants should not infer a lack of evidence but rather a lack of interest and will to explore Aboriginal interaction with the plant communities of Australia' (Pascoe, 2014, p. 54)

This new literature, which draws on not only these accounts but earlier unrecognised scholarship, reveal how the dispossession of Aboriginal people was followed by rapid deterioration of landscapes that they had managed with care for over 40,000 years. Pascoe (ibid., p. 17) writes 'farmers noticed the alarming drop in productivity over a mere handful of years as sheep ate the croplands and compacted the light soils'. Evidence of established structures including grain storage facilities, terracing and intricate fish traps were conveniently omitted from the colonial narrative. Firestick farming, which is today capturing attention as a land management and carbon capture strategy through the work of the Aboriginal Carbon Foundation (http://aboriginalcarbonfoundation.com/), was used to clear areas of land separated by tracts of timber. Like contemporary farmers, Aborigines allocated the poorer soils to forest and used the better soils for cropland (Pascoe, 2014; Gammage, 2012). In regard to sophisticated Aboriginal fish traps, anthropologist Harry Laurandos says, 'the old distinction between "resourceful" agriculturalist and "quiescent" hunter can no longer really be seen to apply'. Research provides evidence of 'active participation' rather than 'passive adaptation' in complex social, environmental and demographic entanglements. In other words, Australian Aborigines were not 'reacting to the state of nature but directly affecting its production' (cited in Pascoe, 2014, p. 50). Further, they were doing so within a highly organised social framework:

> It's difficult to look at the decision-making processes involved in the creation of Aboriginal and Torres Strait Island government and not think of the word democracy ... not all who became old were included in the final decision-making processes ... their role was codified by levels of initiation ... their election was gradual and convoluted, but they didn't assume their position as a result of force or inheritance. They earned the respect of their fellows.
>
> (Pascoe, 2014, p. 131)

Discrediting evidence of permanent settlement and suppressing public disclosure of a complex society not only produced historically distorted information about Aboriginal communities but shutdown opportunity for challenges to the 'morality and legality of the colonial dispossession' (Gerritsen, cited in Pascoe, 2014, p. 83). Eliding the existence of complex building techniques and construction skills makes it easier to dismiss Aboriginal claims to land. This is a strategic move that serves the colonial project and continues to mask dispossession. The new attention to indigenous foods and land management strategies is, like agroecology, susceptible to cooptation. Pascoe welcomes interest in Aboriginal land management and cultivation as 'good news' with a caveat. In reference to recent trials on wild rice, he writes

> [I hope] Aboriginal people will be able to take advantage of the intellectual property they have invested in this plant ... Sun Rice Pty Ltd assisted in the research and, as they stand to benefit from research results, I hope they'll respect the contribution Aboriginal people made to the research over thousands of years.
>
> (Pascoe, 2014, p. 38)

Christopher Mayes (2018) alerts us to the dangers in de-subjugating and then recolonising these knowledges in emerging discourses such as food sovereignty. His aptly named *Unsettling Food Politics: Agriculture, Dispossession and Sovereignty in Australia* (2018) explores the possibilities for 'ethical and political interactions with food' (ibid., p. 11) in a country where many would prefer to forget 'the blood spilled in the past, which has allowed for benefits to be gained and enjoyed in the present' (Patel cited in Mayes, 2018). Mayes refers to discursive efforts to counter the organised amnesia over dispossession imposed by White Australia such as the Welcome to Country (WTC) and Acknowledgement of Country (AOC) rituals commonly performed at public events. He remarks:

> ... even with these rituals commonly performed, Indigenous issues, peoples and histories are soon 'disremembered'. The WTC and AOC practices are vulnerable to being swallowed by the silence of disremembrance that functions as a remembering to forget. That is, the

> brief acknowledgement of dispossession and recognition of Indigenous belonging allows a non-indigenous audience to forget the uncomfortable history and move on with its event.
>
> (Mayes, 2018, p. 142)

Mayes seeks to remind us that such speech acts are examples of our own performance of identity – efforts to voice our own 'anti-racist credentials' – but, more importantly, also risk contributing to the 'terra nullius mythology' (Hart cited in Mayes, 2018, p. 143) that First Nations people exist in some romantic fiction.

How do we 'embrace a return of knowledge in ways that don't simply use those knowledges as a means to reinforce white settler belonging' (Mayes, 2018, p. 143)? Voice is not read transparently and power shapes the emergence of voice under particular conditions. Further, voice is dialogue and requires an *active listener*. Megan Davis, a Cobble Cobble woman from Queensland and professor of law at University of New South Wales, says *The Uluru Statement from the Heart* (2017), an appeal for indigenous cultural recognition, presents 'a moral challenge to all Australians: hear our voices, and pause to listen and understand' (cited in Mayes, 2018, p. 167). The telling of one's story, one's own account, in public and being heard is vital to recognition but only the valuing and accurate reading of that account is a '*possible* starting point for recognising someone as a political subject' (Couldry, 2010, p. 109). In this respect a shared vision for the future is not possible 'until the past is faced objectively and communion is made with our errant history' (Hawken, 2007, p. 188). Only listening can help us 'remember rightly and hear the voices that disturb the silence and unsettle the comfortable and relaxed stories we tell ourselves' (Mayes, 2018, p. 139).

Conclusion

In this book agroecology and other forms of learning have been presented as resistive epistemologies that draw on a variety of conceptions 'of what it means to know, of what counts as knowledge and how that knowledge is produced' (Santos, 2007, p. xxi). Given 'all social practices involve knowledge [and] the production of knowledge is, in itself, a social practice' which is self-reflexive and shapes practices in motive and actions, proposals for radicalising democracy and decolonising knowledge and power are 'only feasible if the dominant epistemology is subject to a critique allowing for the emergence of epistemological options that give credibility to the forms of knowledge that underlie those proposals' (ibid.). This demands a reconsideration of value foundations that are constantly reaffirmed under neo-liberalism. Decolonisation provides a way to challenge these foundations both epistemologically and ontologically involving, as it does, 'value-relationships, approaches to rights and a shift from an economic to an ecological calculus that concurrently challenges the rules and relations of a corporate or neo-liberal food regime' (Burch, 2017, p. 139).

The original meaning of crisis is decision-point. As our food system becomes more and more toxic to our collective well-being we must look to examples of doing food that privilege purpose and people over profit. We must insist on forms of organising foodways that embody social cooperation and voice as any other type of system that seeks to provide us with food is illegitimate and blind to everyone's needs. We need to rethink our identities as eaters, not consumers; reintegrate the voices lost in the food system, including those without citizenship, as part of the broader polity; and restore respect through genuine social inclusion and participation in democratic life. Social change is 'less about *fighting against* an essentially abstract and unknowable "global industrial food system", than it is about *fighting for* highly pragmatic interests, such as certain desired ways of planting, procuring, cooking, or eating' (Sherwood et al., 2017a, p. 199). The realities of people's lives on the streets and around their tables, and the connections that emerge through practice and voice, are empirical realities that are frequently left out of oppositional narratives pitching small-scale farmers against agribusiness. Risks of co-optation, reductionism and replication are mitigated by diverse processes of production and exchange that incorporate voices that are left out of the conventional food system.

The next big thing is lots of these small things united by a common hunger for change. Initiatives that 'emerge from the awareness of being small pieces of the bigger jigsaw puzzle' (González, 2017, p. 155) that is food bring an essentially local character to global struggles. Knowledge includes not only cognitive knowledges but 'the affects, experiences and skills gained through life, stored in the physical landscape and built into the environment and social interactions' (Sherwood et al., 2017b, p. 217). It is 'people's public understanding of their social categories, such as ideas, relationships and effects of relative, friends, neighbours and institutions, but also their practice' (ibid.). Multiple knowledges and peer-to-peer knowledge sharing provide opportunities to push back against asymmetrical power structures. Further, 'creative forms of governance and exchange give people a new identity as members of connected networks and an avenue for collective action, beyond typical classifications that label them as consumers or growers' (González, 2017, p. 155).

This diversity is key to understanding the *unconventional* solution to transformative change in food systems. It is a conglomeration of allies 'across and beyond food movements coming together under a solidarity economy model able to create solid alternatives for regenerative ways of eating and living that cannot and should not be conventionalised' (ibid.). Resistance against corporate capture is being led by transnational food movements such as LVC on the global stage. These efforts are being reinforced by local food movements 'working in customary and contemporary alter- and counter-hegemonic niches in order to build a "globalisation from below"' (Vivero-Pol, 2017, p. 17). Entreating us to 'eat locally, but reclaim globally', José Vivero-Pol well argues that if our norms around food shape our priorities for action – our political attitudes and beliefs and, by extension, the policies we support – consideration of food as commons will open up spaces for policy change and claims in the future.

This calls for a 'reconceptualization of power' (Santos, 2007, p. xxxviii) that reassesses not only colonial relations but existing political structures. 'For there to be food justice there must be food sovereignty. For there to be food sovereignty there must be land justice. For there to be land justice there must be a commitment to decolonisation' (Sbicca, 2018, p. 1100; Kerssen & Brent, 2017; Grey & Patel, 2014). Justice in terms of food is always co-constitutive with other forms of justice (Herman et al., 2018), just as 'different forms of oppression or domination generate equally distinct forms of collective resistance, mobilization, subjectivity, and identity, which invoke differentiated notions of justice and dignity' (Santos, 2007, p. xivi). Universals travel globally but ideals like human rights, justice and sovereignty are empty signifiers until attached to people through place. Further, they are not static. Grace Lee Boggs urges us to 'break free from views that were at one time liberating but had become fetters because reality had changed' (Boggs, 2012, p. 58).

Our local perspectives must be informed by analyses of growing corporate power and financialisation of food systems. International food systems are multi-scalar so broader interconnections between the food justice and more radical food sovereignty movements are essential to normalising real values and avoiding the depoliticisation of discourses by Big Food. Struggles against the appropriation and outright theft of land and biodiversity must be incorporated into our analysis and activism in counter-hegemonic public spheres that support a 'subaltern cosmopolitanism' that recognise 'a plurality of knowledges and distinctive conceptions of human dignity' (Santos, 2007, p. xivii). Key to this goal is addressing the power and privilege in movements as well as the wider communities in which we live and eat. Holt-Giménez refers to 'gatekeepers of the dominant food movement discourse' (2017, p. 173) many of whom are white, male professionals, academics and bureaucrats. A radical leadership comprised of women, people of colour, peasants, farmers and food workers with the skills and knowledge central to individual and community capacity-building is needed to take the not-so-quiet steps towards a transformed food future.

That future must include everyone who eats.

References

ABC News. (2018, August 11). Monsanto ordered to pay $289 million to man who claims Roundup weed killer caused cancer. Retrieved from the *ABC News* website www.abc.net.au/news/2018-08-11/monsanto-ordered-to-pay-289-million-to-school-groundskeeper/10109352

AFP. (2016, September 15). *Bayer–Monsanto merger an $88b 'marriage made in hell'*. Retrieved from www.news.com.au/finance/business/manufacturing/bayer-monsanto-merger-an-88b-marriage-made-in-hell/news-story/23e17cf89cbf1a98a6c413410a6afd25

Alberti, M., Marzluff, J. M., Shulenberger, E., Bradley, G., Ryan, C. & Zumbrunnen, C. (2003). Integrating humans into ecology: Opportunities and challenges for studying urban ecosystems. *BioScience*, *53*(12), 1169–1179.

Alcazar, L., Rabar, M., Lopez, K., Markham, C. & Sharma, S. (2017). Examining the impact of a school-based fruit and vegetable co-op in the Hispanic community through documentary photography. *Appetite*, *116*, 115–122.

Alkon, A. & Agyeman, J. (2011). *Cultivating food justice: Race, class, and sustainability*. Cambridge, MA: MIT Press.

Alkon, A. & Guthman, J. (2017). *The new food activism: Opposition, cooperation and collective action*. Oakland, CA: University of California Press.

Althusser, L. (1970). Ideology and ideological state apparatuses: notes towards an investigation'. In L. Althussar (Ed.), *Lenin and philosophy and other essays*, Monthly review press.

Altieri, M. A. (1999). Applying agroecology to enhance productivity of peasant farming systems in Latin America. *Environment, Development and Sustainability*, *1*, 197–217.

Altieri, M. A. (2002). Agroecology: The science of natural resource management for poor farmers in marginal environments. *Agriculture, Ecosystems and Environment*, *93*, 1–24.

Altieri, M. A. & Nicholls, C. I. (2017). Agroecology: A brief account of its origins and currents of thought in Latin America. *Agriculture and Sustainable Food Systems, 3–4*, 231–237.

Altieri, M. A. & Toledo, V. M. (2011). The agroecological revolution in Latin America: Rescuing nature, ensuring food sovereignty and empowering peasants. *Journal of Peasant Studies*, *38*(3), 587–612. doi:10.1080/03066150.2011.582947

Amin, A. (2006). The good city. *Urban Studies*, *43*(5–6), 1009–1023.

Andrée, P., Ayres, J., Bosia, M. J. & Massicotte, M.-J. (Eds.). (2014). *Globalization and food sovereignty: Global and local change in the new politics of food.* Toronto: University of Toronto Press.

Anguelovski, I., Connally, J. T., Masip, L. & Persall, H. (2017). Assessing green gentrification in historically disenfranchised neighbourhoods: A longitudinal and spatial analysis of Barcelona. *Urban Geography*, *39*(3), 458–491.

Arendt, H. (1958). *The human condition.* Chicago: Chicago University Press.

Armitage, D. & Plummer, R. (Eds.). (2010). *Adaptive capacity and environmental governance.* Berlin: Springer-Verlag.

Ashcraft, K. L., Kuhn, T. R. & Cooren, F. (2009). Constitutional amendments: 'Materializing' organizational communication. In J. P. Walsh & A. P. Brief (Eds.), *The Academy of Management Annals, Vol. 3* (pp. 1–64). London: Routledge.

Augé, M. (1995). *Non-places: Introduction to an anthropology of supermodernity* (J. Howe, Trans.). New York: Verso.

Australian Institute of Agricultural Science and Technology. (2016). Biotechnology and agricultural innovation policy paper. *Agricultural Science*, 2(16), 66.

Baiocchi, G. & Ganuza, E. (2015). Democratising inequalities: Dilemmas of the new public participation. In C. Lee, M. McQuarries & E. T. Walker (Eds.), *Becoming a best practice: Neoliberalism and the curious case of participatory budgeting* (Vol. 187–203). New York: New York University Press.

Balibar, É. (2004). *We, the people of Europe?* Princeton: Princeton University Press.

Barber, B. (2004). *Strong democracy: Participatory politics for a new age.* Berkeley: University of California Press.

Barking and Dagenham Growth Commission. (2015). *No one left behind: In pursuit of growth for the benefit of everyone.* Retrieved from the London Borough of Barking & Dagenham website www.lbbd.gov.uk/sites/default/files/attachments/No-one-left-behind-in-pursuit-of-growth-for-the-benefit-of-everyone.pdf

Barnhill, A. & Doggett, T. (2018). Food ethics 1: Food production and food justice. *Philosophy Compass*, *13*, 1–13.

Bayer. (2018a). *Transparency initiative by Bayer wins best public outreach program at Agrow Awards 2018.* Retrieved from https://media.bayer.com/baynews/baynews.nsf/id/Transparency-Initiative-by-Bayer-wins-Best-Public-Outreach-Program-at-Agrow-Awards-2018

Bayer. (2018b). *Crop science: Understanding regulatory science for crop protection.* Retrieved from www.cropscience-transparency.bayer.com/Assets/BCS-INTER/WS_Prisma/tutorials/04/en/index.html

Bayer AG & The Monsanto Company. (2018, 20 November). *Advancing Together as One.* Retrieved from www.bayer.com/en/advancing-together.aspx

BBC News. (2018a, December 10). France yellow vest protests: Macron promises wage rise. *BBC News.* Retrieved from www.bbc.com/news/world-europe-46513189

BBC News. (2018b, December 18). France protests: The voices of the 'gilets jaunes'. *BBC.* Retrieved from www.bbc.com/news/world-europe-46480867

BCFN, MUFPP. (2018). *Food & cities: The role of cities for achieving the Sustainable Development Goals.* Retrieved from the Barilla Centre for Food & Nutrition website www.barillacfn.com/media/material/food_cities.pdf

Belasco, W. (2012). *Meals to come: A history of the future of food.* Berkeley: University of California Press.

Benhabib, S. (Ed.). (1996). *Democracy and difference: Contesting the boundaries of the political*. Princeton, NJ: Princeton University Press.

Bimber, B., Flanagin, A. J. & Stohl, C. (2012). *Collective action in organisations: Interaction and engagement in an era of technological change*. Cambridge: Cambridge University Press.

Bittman, M. (2014, June 25). Rethinking the word 'foodie'. *New York Times*. Retrieved from www.nytimes.com/2014/06/25/opinion/mark-bittman-rethinking-the-word-foodie.html

Blanco, G., Gastel, J. v. & Lagarrigue, A. (2017). Assembling responsible food markets: the case of Cooperative La Manzana in southern Chile. In S. Sherwood, A. Arce, & M. Paredes (Eds.), *Food agriculture and social change: The everyday vitality of Latin America* (Vol. 184–197). Abingdon, Oxford: Routledge.

Blay-Palmer, A. (2008). *Food fears: From industrial to sustainable food systems*. Aldershot: Ashgate Publishing.

Blythe, R. (1979). *The look in winter*. London: Allen Lane.

Boff, L. (1994). Social ecology: Poverty and misery. In D. G. Hallman (Ed.), *Ecotheology: Voices from South and North* (pp. 235–247). Maryknoll, NY: Orbis Books.

Boggs, G. L. (2012). *The next American revolution: Sustainable activism for the twenty-first Century*. Berkeley: University of California Press.

Bohstedt, J. (2014). Food riots and the politics of provisions in world history. *Institute of Development Studies Working Paper, 2014*(444), 1–31.

Bolin, T. (2017). Struggling for democracy: Paulo Freire and transforming society through education. *Policy Futures in Education, 15*(6), 744–766.

Booth, S. & Coveney, J. (2015). *Food democracy: From consumer to food citizen*. Basel, Switzerland: Springer.

Born, B. & Purcell, M. (2006). Avoiding the local trap: Scale and food systems in planning research. *Journal of Planning Education and Research, 26*(2), 195–207.

Borras, J. S. M., Edelman, M. & Kay, C. (2008). Transnational agrarian movements: Origins and politics, campaigns and impact. *Journal of Agrarian Change, 8*(2–3), 169–204. doi:10.1111/ j.1471-0366.2008.00167.x

Boulder Statement (1997). Boulder, Colorado: 9th MacBride Roundtable.

Bourdieu, P. (1984). *Distinction: A social critique of the judgement of taste*. Cambridge, MA: Harvard University Press.

Bové, J. & Dufour, F. (2002). *Food for the future: Agriculture for a global age*. Cambridge: Polity Press.

Boyd, D. (2010). Social network sites as networked publics: Affordances, dynamics and implications. In Z. Papacharissi (Ed.), *Networked self: Identity, community, and culture on social network sites* (pp. 39–58). New York: Routledge.

Breindl, Y. (2012). The dynamics of participation and organisation in European digital rights campaigning. *Journal of Democracy, 42*, 24–44.

Brown, P. & Vidal, J. (1999, August 25). GM investors told to sell their shares. *Guardian*. Retrieved from www.theguardian.com/science/1999/aug/25/gm.food

Brown, V. (2007). *Leonardo's Vision: A guide to collective thinking and action*. Boston, MA: Sense Publishers.

Brown, W. (2003). Neo-liberalism and the end of liberal democracy. *Theory & Event, 7*(1), 1–23.

Bruno, K. & Greer, J. (1996). *Greenwash: The reality behind corporate environmentalism*. Penang, Malaysia: Third World Network.

Bullock, J. M., Dhanjal-Adams, K. L., Milne, A., Oliver, T. H., Todman, L. C., Whitmore, A. P. & Pywell, R. F. (2017). Resilience and food security: Rethinking an ecological concept. *Journal of Ecology*, *105*(4), 880–888.

Burch, S. (2017). Everything has a story: Decolonisation, First Nations sovereignty, and the seventh pillar of food sovereignty in the Australian context. In O. Vodeb (Ed.), *Food democracy: Critical lessons in food, communication, design and art* (pp. 127–167). Bristol, UK: Intellect.

Burgess, J. & Matamoros-Fernández, A. (2016). Mapping sociocultural controversies across digital media platforms: One week of #gamergate on Twitter, YouTube, and Tumblr. *Communication Research and Practice*, *2*(1), 79–96.

Butler, J. (1993). *Bodies that matter*. New York: Routledge.

Butler, J. (2005). *Giving an account of oneself*. New York: Fordham University Press.

Cadieux, K. V. & Slocum, R. (2015). What does it mean to do food justice? *Journal of Political Ecology*, *22*, 1–25.

Cadigan, H. (2018, November 7). Alexandria Ocasio-Cortez learned her most important lessons from restaurants. Retrieved from the *bon appétit* website www.bonappetit.com/story/alexandria-ocasio-cortez-lessons-from-restaurants

Caldwell, M. (2018). Hacking the food system: Re-making technologies of food justice. In T. Schneider, K. Eli, C. Dolan & S. Ulijaszek (Eds.), *Digital food activism* (pp. 25–42). Abingdon, Oxfordshire: Routledge.

Callon, M., Lascoumes, P. & Barthe, Y. (2009). *Acting in an uncertain world. An essay on technical democracy* (G. Burchell, Trans.). Cambridge, MA: MIT Press.

Calvário, R. & Kallis, G. (2016). Alternative food economies and transformative politics in times of crisis: Insights for the Basque Country and Greece. *Antipode*, 1–20.

Campanella, T. J. (2006). Urban Resilience and the Recovery of New Orleans. *Journal of the American Planning Association*, *72*(2), 141–146.

Campbell, H., Evans, D. & Murcott, A. (2017). Measurability, austerity and edibility: Introducing waste into food regime theory. *Journal of Rural Studies*, *51*, 168–177.

Campbell, M. C. (2004). Building a common table: The role for planning in community food systems. *Journal of Planning, Education and Research*, *23*(4), 341–355.

Canclini, N. G. (2001). *Consumers and citizens: Globalisation and multicultural conflicts*. Minneapolis: University of Minnesota Press.

Capra, F. (2002). *Hidden connections: Integrating the biological, cognitive, and social dimensions of life into a science of sustainability*. New York: Random House.

Carlson, J. & Chappell, M. J. (2015). *Deepening Food Democracy: The tools to create a sustainable, food secure and food sovereign future are already here – deep democratic approaches can show us how*. Retrieved from www.iatp.org/sites/default/files/2015_01_06_Agrodemocracy_JC_JC_f_0.pdf

Carter, M. (2015). Social inequality, agrarian reform, and democracy in Brazil. In M. Carter (Ed.), *Challenging social inequality: The landless rural workers movement and agrarian reform in Brazil* (p. 7). Durham, NC: Duke University Press.

Caslick, R. (2018). Editor's Letter. *Two Good Times*, 2.

Cavarero, A. (2000). *Relating narratives*. London: Routledge.

CFS. (2012). *Committee on World Food Security*. Retrieved from www.fao.org/cfs/cfs-home/en/

Chapin, F. S., Lovecraft, A. L., Zavalata, E. S., Nelson, J., Robards, M. D., Kofinas, G. P., Trainor, S. F., Peterson, G. D., Huntingdon, H. P. & Naylor, R. L. (2006). Policy strategies to address sustainability of Alaskan Boreal forests in response to a directionally changing climate. *Proceedings of the National Academy of Sciences of the United States of America*, *103*, 16637–16643.

Chappell, M. J. (2018). *Beginning to end hunger: Food and the environment in Belo Horizonte, Brazil, and beyond*. Oakland, CA: University of California Press.

Chappell, M. J. & Schneider, M. L. (2016). The new three-legged stool: Agroecology, food sovereignty, and food justice. In M. Rawlinson & C. Ward (Eds.), *The Routledge Handbook of Food Ethics* (Chapter 39). London: Routledge.

Charão-Marques, F., Schmitt, C. J. & Oliveira, D. (2017). Unfolding agencies and associations of agroecology networks in Brazil. In S. Sherwood, A. Arce & M. Paredes (Eds.), *Food, agriculture and social change: The everyday vitality of Latin America* (Vol. 126–140). Abingdon, Oxford: Routledge.

Chemnitz, C., Luig, B. & Schimpf. M. (Eds.). (2017). *Agri-food atlas: Facts and figures about the corporations that control what we eat*. Retrieved from https://th.boell.org/sites/default/files/agrifoodatlas2017_facts-and-figures-about-the-corporations-that-control-what-we-eat.pdf

Chiseri-Strater, E. (1996). Turning in upon ourselves: Positionality, subjectivity and reflexivity in case study and ethnographic research. In P. Mortenson & G. E. Kirsch (Eds.), *Ethics and representation in qualitative studies of literacy* (pp. 115–132). Illinois: National Council of Teachers of English.

Chohan, J. K. (2017). *Reclaiming the food system: Agroecological pedagogy and the IALA Marìa Cano*. Retrieved from www.alternautas.net/blog/2017/8/17/reclaiming-the-food-system-agroecological-pedagogy-and-the-iala-mara-cano

Chow, L. (2016). *Paris becomes one of the most garden-friendly cities in the world*. Retrieved from www.ecowatch.com/paris-urban-gardening-2039785049.html

Cid Aguayo, B. E. & Latta, A. (2015). Agro-ecology and food sovereignty movements in Chile: Sociospatial practices for alternative peasant futures. *Annals of the Association of American Geographers*, *105*(2), 397–406.

City of Sydney. (2016). *Community Wellbeing Indicators*. Retrieved from www.cityofsydney.nsw.gov.au/__data/assets/pdf_file/0008/264887/Community-Wellbeing-Indicators-Report-2016.pdf

Civic Systems Lab (2017). *Designed to scale: Mass participation to build resilient neighbourhoods*. Retrieved from www.participatorycity.org/report-the-research

Clancy, K. (2004). Potential contributions of planning to community food systems. *Journal of Planning Education and Research*, *23*(4), 435–438.

Clapp, J. (2014). Financialization, distance and global food politics. *Journal of Peasant Studies*, *41*(5), 797–814.

Clark, S. (2013). Climate Change Will Force Farmers to Adapt. Retrieved from the *ABC News* website www.abc.net.au/news/2013-01-15/climate-change-will-force-farmers-to-adapt-csiro-says/4464506

Cloke, P., May, J. & Williams, A. (2017). The geographies of food banks in the meantime. *Progress in Human Geography*, *41*(6), 703–726.

Cohen, N. & Ilieva, R. T. (2015). Transitioning the food system: A strategic practice management approach for cities. *Environmental Innovation and Societal Transitions*, *17*, 199–217.

Colasanti, K. J. A., Hamm, M. W. & Litjens, C. M. (2013). The city as an 'agricultural powerhouse'? Perspectives on expanding urban agriculture from Detroit, Michigan. *Urban Geography*, *33*(3), 348–369.

Coleman, R. (2017). Soil currency: Exploring a more equitable, sustainable, and participatory economic system. In J. Duncan & M. Bailey (Eds.), *Sustainable food futures: Multidisciplinary solutions* (pp. 175–185). London: Routledge.

Connell, R. (2007). *Southern theory: The global dynamics of knowledge in social science*. Cambridge, UK: Polity.

Contesse, M., Vliet, B. J. M. v. & Lenhart, J. (2018). Is urban agriculture urban green space? A comparison of policy arrangements for urban green space and urban agriculture in Santiago de Chile. *Land Use Policy*, *71*, 566–577.

Conversi, D. (2016). Sovereignty in a changing world: From Westphalia to food sovereignty. *Globalizations*, *13*(4), 484–498.

Cooley, A. (2010). Political pedagogy towards democratic education: educating students to care about local and global injustice. *Journal of Thought, Fall–Winter*, 59–61.

Cooren, F., Kuhn, T., Cornelissen, J. P. & Clark, T. (2011). Communication, organizing and organization: An overview and introduction to the special issue. *Organization Studies*, *32*(9), 1149–1170. doi:http://dro.dur.ac.uk/8259/1/8259.pdf?DDD2

Copeland, L. (2016). Research frontiers in agrifood systems. *Agricultural Science*, *2*(16), 6–12.

Cottle, S. & Lester, L. (2011). *Transnational protest and the media*. New York: Peter Lang.

Couldry, N. (2010). *Why voice matters: Culture and politics after neoliberalism*. London: Sage.

Cox, A. M. (2005). What are communities of practice? A comparative review of four seminal works. *Journal of Information Science*, *31*(6), 527–540.

Craven, L., Scholsberg, D. & Mann, A. (2018). Addressing the SDGs through food business incubation: FoodLab Sydney. *Food & cities: The role of cities for achieving the sustainable development goals*, 145–151.

Curiel, C. (2017). Deepening relationships through bio-intensive food. In S. Sherwood, A. Arce & M. Paredes (Eds.), *Food agriculture and social change: The everyday vitality of Latin America* (Vol. 141–152). Abingdon, Oxford: Routledge.

Czarniawska, B. (2013). Organisations as obstacles to organising. In D. Robichaud & F. Cooren (Eds.), *Organisation and organising. Materiality, agency and discourse* (pp. 3–22). New York, NY: Routledge.

Darnhofer, I. (2010). Strategies of family farms to strengthen their resilience. *Environmental Policy Governance*, *20*(4): 212–222.

da Silva, J. (2012). Better governance to achieve food security. *Inter Press Service News*. Retrieved from www.ipsnews.net/2012/11/better-governance-to-achieve-food-security

Davies, A. R. & Legg, R. (2018). Fare sharing: interrogating the nexus of ICT, urban food sharing, and sustainability. *Food, Culture & Society*, *21*(2), 233–254. doi:10.1080/15528014.2018.1427924

Davis, G., McAdam, D., Scott, W. R. & Zald, M. N. (2005). *Social movements and organisation theory*. Cambridge: Cambridge University Press.

De Angelis, M. (2017). *Omnia sunt communia: On the Commons and the transformation to postcapitalism*. Chicago, Il.: The University of Chicago Press.

de Certeau, M. (1984). *The practice of everyday life*. Berkeley: University of California Press.

Declaration of Nyéléni. (2007). Retrieved from the Nyéléni website http://nyeleni.org/spip.php?article290

De Cunto, A., Tegoni, C., Sonnino, R., Michel, C. & Lajili-Djalaï, F. (2017). *Food in cities: Study on innovation for a sustainable and healthy production, delivery, and consumption of food in cities*. Retrieved from the European Commission website https://ec.europa.eu/research/openvision/pdf/rise/food_in_cities.pdf

Deetz, S. & Mumby, D. (1990). Power, discourse, and the workplace: Reclaiming the critical tradition. In J. A. Anderson (Ed.), *Communication yearbook 13* (pp. 18–47). London: Sage.

Defra. (2014). *Food statistics pocket book* (in year update), Retrieved from https://assets.publishing.service.gov.uk/government/uploads/system/uploads/attachment_data/file/423616/foodpocketbook-2014report-23apr15.pdf

Delegates to the Food and Agriculture Organisation (FAO) on Plant Genetic Resources. (1998). *Let Nature's Harvest Continue!: African Counter-Statement to Monsanto*. Retrieved from www.ukabc.org/gaiam2_2.htm

della Porta, D. (2012). Critical trust: Social Movements and democracy in times of crisis. *Anno II, 4 December*, 33–44.

della Porta, D. (2013). *Can democracy be saved?* Cambridge: Polity.

Demailly, K.-E. & Darly, S. (2017). Urban agriculture on the move in Paris: The routes of temporary gardening in the neoliberal city. *Acme*, *16*(2), 332–361.

de Paz García, J. (2018). *Three Madrid-based collectives design a self-sufficient public infrastructure*. Retrieved from the *domus* website www.domusweb.it/en/architecture/2018/04/04/three-madrid-based-collectives-design-a-self-sufficient-public-infrastructure.html

de Schutter, O. (2017). The political economy of food systems reform. *European Review of Agricultural Economics*, *44*(3), 705–731.

Desmarais, A. A. (2007). *La Vía Campesina: Globalization and the power of peasants*. London: Pluto Press.

Desmarais, A. A. & Nicholson, P. (2013). La Vía Campesina: An historical and political analysis. In *La Vía Campesina's open book: Celebrating 20 years of struggle and hope*. Retrieved from https://viacampesina.org/en/la-via-campesina-s-open-book-celebrating-20-years-of-struggle-and-hope/

de Souza Briggs, X. (2008). *Democracy as problem-solving: Civic capacity in communities across the globe*. Cambridge: The MIT Press.

Dewey, J. (1916). *Education and democracy*. Mineola, NY: Dover Publications Inc.

Diani, M. (2003). 'Leaders' or brokers? Positions and influence in social movement networks. In M. Diani & D. McAdam (Eds.), *Social movements and networks: Relational approaches to collective action* (pp. 105–122). Oxford: Oxford University Press.

Dixon, B. A. (2014). Learning to see food justice. *Agriculture and Human Values*, *31*(2), 175–184.

Dixon, J., Beilin, R., Isaacs, B., Rose, N., Santhanam-Martin, M. & Young, M. (2011). *Food security for the short or long term? Australia and New Zealand Agri-Food Research Network*. Retrieved from https://afsa.org.au/blog/2011/06/06/ood-security_short_long_term/

Dowling, R., Pauline, M. & Bulkeley, H. (2014). Retrofitting cities: Local governance in Sydney. *Australia. Australian Cities*, *38*, 18–24.

Drouet, L. & Lacrouts, O. (2018, May 4). *Paris. The Agrocité is back*. Paris: domus.

Duncan, J. A. B. & Barling, D. (2012). Renewal through participation in global food security governance: Implementing the international food security and nutrition civil society mechanism to the committee on world food security. *International Journal of Sociology of Agriculture and Food*, *19*(2), 143–161.

Duvernoy, I. (2018). Alternative voices in building a local food policy: Forms of cooperation between civil society organisations and public authorities in and around Toulouse. *Land Use Policy*, *75*, 612–619.

Eakin, H., Bohle, H. G., Izac, A. M., Reenberg, A., Gregory, P. & Pereira, L. (2010). Food, Violence and Human Rights. In P. E. John Ingram, Diana Liverman (Ed.), *Food security and global environmental change* (pp. 245–271). London: Earthscan.

Edelman, M. (2014). Food sovereignty: Forgotten genealogies and future regulatory challenges. *Journal of Peasant Studies*, *41*(6), 959–978.

Edwards, F. & Davies, A. (2018). Connective consumptions: Mapping Melbourne's food sharing ecosystem. *Urban Policy and Research*. Retrieved from www.tandfonline.com/doi/abs/10.1080/08111146.2018.1476231

Ehrmann, S. (2018, November 29). *Green gentrification and how to avoid it*. Foreground.

Elver, H. (2018). *Report of the special rapporteur for the right to food* (A/HRC/34/48). Retrieved from https://documents-dds-ny.un.org/doc/UNDOC/GEN/G17/017/85/PDF/G1701785.pdf?OpenElement

Escobar, A. (1995). *Encountering Development: The Making and Unmaking of the Third World*. Princeton, NJ: Princeton University Press.

Escobar, A. (1998). Whose knowledge, whose nature? Biodiversity, conservation, and the political ecology of social movements. *Journal of Political Ecology*, *5*(1), 53–82.

Escobar, A. (2008). *Territories of difference: Place, movements, life, redes*. Durham, NC: Duke University Press.

ETC. (2017). *Who will feed us? The industrial food chain vs the peasant food web*. Retrieved from the Action Group on Erosion, Technology and Concentration website www.etcgroup.org/content/who-will-feed-us-industrial-food-chain-vs-peasant-food-web

Evans, P. B. (1996). Government action, social capital and development: Reviewing the evidence on synergy. *World Development*, *24*(6), 1119–1132.

FAO. (2015a). *Final Report for the International Symposium on Agroecology for Food Security and Nutrition*. Retrieved from the United Nations Food & Agriculture website www.fao.org/3/a-i4327e.pdf

FAO. (2015b). *Feeding a Growing World Population Requires Investments in Rural Areas*. Retrieved from Milan www.fao.org/news/story/en/item/337997/icode/

FAO. (2018a). *Catalysing dialogue and cooperation to scale up agroecology: Outcomes of the FAO regional seminars on agroecology*. Retrieved from the United Nations Food and Agriculture website www.fao.org/3/I9035EN/i9035en.pdf

FAO. (2018b). *The Role of Cities in the Transformation of Food Systems: Sharing lessons from the Milan Pact Cities*. Retrieved from www.milanurbanfoodpolicypact.org/wp-content/uploads/2018/10/CA0912EN.pdf

Feenstra, G. (2002). Creating space for sustainable food systems: Lessons from the field. *Agriculture and Human Values*, *19*(2), 99–106.

Fernandes, M. B. (2015). The formation and territorialisation of the MST in Brazil. In M. Carter (Ed.), *Challenging social inequality: The landless rural workers movement and agrarian reform in Brazil* (pp. 115–148). Durham, NC: Duke University Press.

Figueroa, M. (2015). Food Sovereignty in Everyday Life: Toward a People-centered Approach to Food Systems. *Globalizations*, *12*(4), 498–512.

Figueroa, M. & Alkon, A. H. (2017). Cooperative social practices, self-determination, and the struggle for food justice in Oakland and Chicago. In A. H. Alkon & J. Guthman (Eds.), *The new food activism: Opposition, cooperation and collective action* (pp. 206–231). Oakland, CA: University of California Press.

Filardi, M. E. & Prato, S. (2018). Reclaiming the future of food: Challenging the dematerialization of food systems. *Right to Food and Nutrition Watch: When Food Becomes Immaterial: Confronting the Digital Age*, *10*, 6–13.

Fisher, F. (2000). *Citizens, experts and the environment*. Durham, NC: Duke University Press.

Flannery, T. (2010). *Here on earth*. Melbourne: Text Publishing.

Fligstein, N. & McAdam, D. (2012). *A theory of fields*. Oxford: Oxford University Press.

Folke, C. (2006). Resilience: The emergence of a perspective for socio-ecological systems analyses. *Global Environmental Change*, *16*(3), 253–67.

Foodbank. (2018). *Foodbank hunger report 2018*. Retrieved from the Foodbank website www.foodbank.org.au/wp-content/uploads/2018/10/2018-Foodbank-Hunger-Report.pdf

Food Ethics Council. (2017). *Food citizenship: How thinking of ourselves differently can change the future of our food system*. Retrieved from www.foodethicscouncil.org/uploads/publications/FoodCitizenship2017_web.pdf

Foodworks. (2010). *FoodWorks: A vision to improve NYC*. Retrieved from http://growingfoodconnections.org/gfc-policy/foodworks-a-vision-to-improve-nycs-food-system/

Fraley, R. (1992). Sustaining the food supply. *Bio/Technology*, *10*(1) January, 40–43.

Fraser, N. (1999). Social justice in the age of identity politics: Redistribution, recognition, and participation. In Ray, L. & Sayer, R. (Eds.), *Culture and economy after the cultural turn* (pp. 25–52). London: Sage.

Fraser, N. (2000). Rethinking recognition. *New Left Review*, *3*, 107–120.

Fraser, N. (2005). Reframing global justice. *New Left Review*, *36*, 69–90.

Fraser, N. (2007). Transnationalizing the public sphere. *Theory, Culture & Society*, *24*(4), 7–30.

Fraser, N. (2008). *Scales of justice: Reimagining political space in a globalising world*. Cambridge: Polity Press.

Freedgood, J. F., Pierce-Quinonez, M. P. & Meter, K. M. (2011). Emerging assessment tools to inform food system planning. *Journal of Agriculture, Food Systems, and Community Development*, *2*(1), 53–64.

Freire, P. (1970). *Pedagogy of the oppressed*. NY: Continuum.

Freire, P. (1998). *Pedagogy of freedom: Ethics, democracy, and civic courage*. Lanham, MD: Rowman and Littlefield.

Friedland, W. (2008). Agency and the agrifood system. In W. Wright & G. Middendorf (Eds.), *The fight over food: Producers, consumers, and activists challenge the global food system*. University Park, PA: The Pennsylvania State University Press.

Fuchs, C. (2006). The self-organization of social movements. *Systemic Practice and Action Research*, *19*(1), 101–137. doi:10.1007/s11213-005-9006-0

Fujisawa, M., Kobayashi, K., Johnston, P. & New, M. (2015). What drives farmers to make top-down or bottom-up adaptation to climate change and fluctuations. A comparative study on three cases on apple farming in Japan and South Africa. *PloS ONE 10*(3): e0120563 doi:10.1371/journal.pone.0120563

Fung A. & Wright, E. O. (2003). *Deepening democracy: Institutional innovations in empowered participatory governance*. Volume IV in The Real Utopia Project. London: Verso.

Galeano, E. (1971). *Open veins of Latin America: Five centuries of the pillage of a continent*. Uruguay: Monthly Review.

Gálvez, A. (2018). *Eating NAFTA: Trade, food polices, and the destruction of Mexico NAFTA*. Oakland, CA: University of California Press.

Gammage, B. (2012). *The biggest estate on earth*. NSW, AU: Allen & Unwin.

Ganesh, S. & Stohl, C. (2010). Qualifying engagement: A study of information and communication technology and the global social justice movement in Aotearoa New Zealand. *Communication Monographs*, *77*(1), 51–74. doi:10.1080/036377 50903514284

Ganesh, S., Zoller, H. & Cheney, G. (2005). Transforming resistance, broadening our boundaries: Critical organizational communication meets globalization from below. *Communication Monographs*, *72*(2), 169–191. doi:10.1080/03637750 500111872

Garrido, S. (2016). *National school of agroecology*. Retrieved from www.transit socialinnovation.eu/sii/ctp/national-school-of-agroecology

George, S. (2015). *Shadow sovereigns: How global corporations are seizing power*. Cambridge: Polity Press.

Ghosh, A. (2016). *The great derangement: Climate change and the unthinkable*. Chicago: The University of Chicago Press.

Gibson, J. (1979). *The ecological approach to visual perception*. Boston: Houghton Mifflin.

Gibson-Graham, J. (1996). *The end of capitalism (as we know it): A feminist critique of political economy*. Minneapolis: University of Minnesota Press.

Gilbert, J. (2008). *Anticapitalism and culture*. Oxford: Berg.

Gilligan, C. (1982). *In a different voice*. Cambridge, MA: Harvard University Press.

Giraldo, O. F. & Rosset, P. M. (2017). Agroecology as a territory in dispute: Between institutionality and social movements. *Journal of Peasant Studies*, *45*(3), 545–564. doi:10.1080/03066150.2017.1353496

Giraud, E. (2018). Displacement, 'failure' and friction: Tactical interventions in the communication ecologies of anti-capitalist food activism. In T. Schneider, K. Eli, C. Dolan & S. Ulijaszek (Eds.), *Digital food activism* (pp. 130–150). Abingdon, Oxfordshire: Routledge.

Giroux, H. A. (2010, 15 December). *Lessons to be learned from Paulo Freire as education is being taken over by the mega rich*. Retrieved from www.truth-out.org/archive/component/k2/item/93016:lessons-to-be-learned-from-paulo-freire-as-education-is-being-taken-over-by-the-mega-rich

Gliessman, S. (2015). *Agroecology: The ecology of sustainable food systems* (3rd ed.). Boulder, CO: CRC.

Gliessman, S. (2016). Editorial: Transforming food systems with agroecology. *Agroecology and Sustainable Food Systems*, *40*(3), 187–189.

Glover, D. (2007). *Monsanto and smallholder farmers: A case-study on corporate accountability*. IDS Working Paper for the DRC on Citizenship, Participation and Accountability. Brighton: University of Sussex Institute of Development Studies.

Glover, D. (2010). The corporate shaping of GM crops as a technology for the poor. *Journal of Peasant Studies*, *37*(1), 67–90.

Golay, C. (2010). The food crisis and food security: Towards a new world food order? *International Development Policy*, *1*, 215–232.

Golding, P. & Murdock, G. (1991). Culture, communication and political economy. In J. Curran & M. Gurevitch (Eds.), *Mass Media and Society* (4 edn, pp. 15–32). London, UK: Edward Arnold.

Gonzalez Novo, M., Castellanos Quintero, A. & Masalis, J. L. P. (Eds.). (2008). *Testimonios: Agricultura Urbana en Cuidad de La Habana*. Havana: Oxfam Cuba and ACTAF.

González, R. A. (2017). The solution cannot be conventionalised: Protecting the alterity of fairer and more sustainable food networks. In J. Duncan & M. Bailey (Eds.), *Sustainable food futures: Multidisciplinary solutions* (pp. 145–160). London: Routledge.

Goodman, D. & Goodman, M. (2001). Sustaining foods: Organic consumption and the socio-ecological imaginary. In M. Choen & J. Murphy (Eds.), *Exploring sustainable consumption: Environmental policy and the social sciences* (pp. 97–119). Amsterdam: Pergamon.

Goodman, D. & Goodman, M. (2007). Localism, livelihoods and the 'post organic': Changing perspectives on alternative food networks in the United States. In D. Maye, L. Holloway & M. Kneafsey (Eds.), *Alternative food geographies: Representation and practices* (pp. 23–38). Oxford: Elsevier Science.

Goodman, M. (2016). Food geographies I: Relational foodscapes and the busy-ness of being more-than-food. *Progress in Human Geography*, *40*(2), 257–266.

Gottlieb, R. & Fisher, A. (1996). Community food security and environmental justice: Searching for a common discourse. *Agriculture and Human Values*, *3*(3), 23–32.

Gottlieb, R. & Joshi, A. (2010). *Food justice*. Boston, MA: MIT Press.

Grant, H. (2017). *Agricultural innovation and sustainability, a message from Hugh Grant*. Retrieved from https://monsanto.com/investors/reports/sustainability-reports/

Grasseni, C. (2017). *Food citizens? Collective food procurement in European cities: Solidarity and diversity, skills and scale*. Retrieved from www.universiteitleiden.nl/en/research/research-projects/social-and-behavioural-sciences/food-citizens-collective-food-procurement-in-european-cities

Greenbank, P. (2003). The role of values in educational research: The case for reflexivity. *British Educational Research Journal*, *26*(6), 791–801.

Greer, A. E., Cross-Denny, B., McCabe, M. & Castrogivanni, B. (2016). Giving economically disadvantaged, minority food panty patrons a voice: Implications for equitable access to sufficient, nutritious food. *Family and Community Health, July–September*, *39*(3), 199–206.

Grey, S. & Patel, R. (2014). Food sovereignty as decolonization: Some contributions from indigenous movements to food system and development politics. *Agriculture and Human Values*, *32*(3), 431–444. doi:10.1007/s10460-014-9548-9

Guardiola-Rivera, O. (2011). *What if Latin America ruled the world? How the South will take the North into the 22nd century*. London: Bloomsbury.
Guthman, J. (2011). Excess consumption or over-production: US farm policy, global warming, and the bizarre attribution of obesity. In R. Peet, P. Robbins & M. Watts (Eds.), *Global Political Ecology* (pp. 51–66). Abingdon, Oxfordshire: Routledge.
Guttal, S. (2018). The changing face of food retail in India. *Right to Food and Nutrition Watch: When Food Becomes Immaterial: Confronting the Digital Age, 10*, 32–39.
Habermas, J. (1989). *The structural transformation of the public sphere*. Cambridge: Polity Press.
Hajer, M. A. (1995). *The politics of environmental discourse: Ecological modernisation and the policy process*. New York, US: Oxford University Press.
Hales, M., Dessibourg, N. & Peña, A. M. (2017). *Global cities 2017: Leaders in a world of disruptive innovation*. Retrieved from www.atkearney.com/global-cities/full-report
Hall, R. (2014). *Food sovereignty: A growing activist and intellectual movement*. Retrieved from the Institute for Poverty, Land and Agrarian Studies website www.plaas.org.za/blog/food-sovereignty-growing-activist-and-intellectual-movement
Hammersley, M. (1993). On the teacher as researcher. In P. Chapman (Ed.), *Educational Research: Volume 1: Current Issues*. London: Sage.
Hansen, T. B. & Verkaaik, O. (2009). Introduction – Urban charisma: On everyday mythologies in the city. *Critique of Anthropology, 29*(1), 5–26.
Hardt, M. & Negri, A. (2017). *Assembly*. Oxford: Oxford University Press.
Harper, A., Alkon, A., Shattuck, A., Holt-Giménez, E. & Lambrick, F. (2009). *Food policy councils: Lessons learned*. Retrieved from https://static1.squarespace.com/static/58916bc7cd0f68047817aae2/t/58e42db615d5db35557b322d/1491348923413/DR21-Food-Policy-Councils-Lessons-Learned-.pdf
Harper, K., Sands, C., Horowitz, D. A., Totman, M., Maitín, M., Rosado, J. S., & Alger, N. (2017). Food justice youth development: Using Photovoice to study urban school food systems. *Local Environment, 22*(7), 791–808.
Harris, E. (2009). Neoliberal subjectivities or a politics of the possible? Reading for difference in alternative food networks. *Area, 41*(1), 55–63.
Harvey, D. (1989). From managerialism to entrepreneurialism: The transformation in urban governance in late capitalism. *Human Geography, 71*(3), 3–17. doi:10.1080/01944363.2017.1322914
Harvey, D. (1996). *Justice, nature and the geography of difference*. Cambridge: Wiley-Blackwell.
Harvey, D. (2008). The right to the city. *New Left Review*, Sept.–Oct. Retrieved from https://newleftreview.org/II/53/david-harvey-the-right-to-the-city
Hassanein, N. (2008). Locating food democracy: Theoretical and practical ingredients. *Journal of Hunger & Environmental Nutrition, 3*(2–3), 286–308.
Hawken, P. (2007). *Blessed unrest*. New York: Viking Press.
Hazan, H. (1994). *Old age: Deconstructions and constructions*. Cambridge, UK: Cambridge University Press.
Heidegger, M. (1971). Building dwelling thinking. In *Poetry, language, thought*. New York: Harper Row.
Hendrickson, M. K., Wilkinson, J., Heffernan, W. D. & Gronski, R. (2008). *The global food system and nodes of power*. Retrieved from http://ssrn.com/abstract=1337273

Herman, A. & Goodman, M. (2018). New spaces of food justice. *Local Environment: The International Journal of Justice and Sustainability*, *23*(11), 1041–1046.

Herman, A., Goodman, M. & Sage, C. (2018). Six questions for food justice. *Local Environment, 23*(11), 1075–1089.

Herod, A. (1999). Reflections on interviewing foreign elites: Praxis, positionality, validity and the cult of the insider. *Geoforum*, *30*(4), 313–327.

Heynen, N., Kurtz, H. & Trauger, A. (2012). Food justice, hunger and the city. *Geography Compass*, *6*(5), 304–311.

Hinrichs, C. & Allen, P. (2008). Selective patronage and social justice: Local food consumer campaigns in historical context. *Journal of Agricultural and Environmental Ethics*, *21*(4), 329–352.

Hislop, R. (2014). *Reaping equity across the USA: FJ organisations observed at the national scale*. (Master's Thesis), University of California-Davis.

Hodgson, C. (2018). *Monopoly and monoculture*. Retrieved from the Medium website https://medium.com/the-food-issue-weapons-of-reason/global-food-industry-monopoly-monoculture-growing-corporate-profits-3df0ae5f7237

Hodgson, J. (1992). Biotechnology: feeding the world? *Bio/Technology*, *10*(1), 47–50.

Holley, K. (2016). *Equitable and inclusive civic engagement: A guide to transformative change*. Columbus: Ohio State University.

Holt-Giménez, E. (2002). Measuring farmers' agroecological resistance after Hurricane Mitch in Nicaragua: A case study in participatory, sustainable land management impact monitoring. *Agriculture, Ecosystems and Environment*, *93*, 87–105.

Holt-Giménez, E. (2009). From food crisis to food sovereignty: The challenge of social movements. *Monthly Review*, *61*(3), 142–156.

Holt-Giménez, E. (2017). *A foodie's guide to capitalism: Understanding the political economy of what we eat*. New York: Monthly Review Press.

Holt-Giménez, E., Bunch, R., Vasquez, J., Wilson, J., Pimbert, M., Boukary, B. & Kneen, C. (2010). Linking farmers' movements for advocacy and practice. *Journal of Peasant Studies*, *37*(1), 203–236. doi:10.1080/03066150903499943

Holt-Giménez, E. & Shattuck, A. (2011). Food crises, food regimes and food movements: Rumblings of reform or tides of transformation? *Journal of Peasant Studies*, *38*(1), 109–144.

Honneth, A. (1995). *The fragmented world of the social*. Albany: SUNY Press.

Honneth, A. (2007). *Disrespect*. Cambridge, UK: Polity Press.

Horst, M., McClintock, N. & Hoey, L. (2017). The intersection of planning, urban agriculture, and food justice: A review of the literature. *Journal of the American Planning Association*, *83*(3), 277–295. doi:10.1080/01944363.2017.1322914

Hospes, O. (2014). Food sovereignty: The debate, the deadlock, and a suggested detour. *Agriculture and Human Values*, *31*, 119–130.

Houston, D., Ghosh, S., Mann, A., McGregor, A., Hsu, J., Chaudhary, S. & Shirozhan, S. (2019). *Growing food and density together: Enabling sustainable urban transformation through local food production*. Sydney: Urban Growth NSW.

Howard, P. (2016). *Concentration and power in the food system: Who controls what we eat?* New York, US: Bloomsbury.

Hugo, G. (2014). *World migration report 2015: Urban migration trends, challenges, responses and policy in the Asia-Pacific*. Retrieved from the International Organisation for Migration website www.iom.int/sites/default/files/WMR-2015-Background-Paper-GHugo.pdf

IAASTD. (2009). *Agriculture at a crossroads*. Retrieved from www.fao.org/fileadmin/templates/est/Investment/Agriculture_at_a_Crossroads_Global_Report_IAASTD.pdf

IAPP. (2016). *Code of Ethics*. Retrieved from the International Association of Public Participation website www.iap2.org.au/About-Us/About-IAP2-Australasia-/Code-of-Ethics

Ingold, T. (2000). *The perception of the environment: Essays in livelihood, dwelling and skill*. London: Routledge.

IPC. (2015). *Declaration of the international forum of agroecology*. Retrieved from the International Planning Committee for Food Sovereignty website www.foodsovereignty.org/forum-agroecology-nyeleni-2015/

IPCC. (2001). *Climate change 2001: Impacts, adaptation and vulnerability*. IPCC Third Assessment Report. Cambridge, MA: Cambridge University Press.

IPCC. (2014). Urban areas. In C. B. Field, V. R. Barros, D. J. Dokken, K. J. Mach, M. D. Mastrandrea, T. E. Bilir, M. Chaterjee, K. L. Ebi, Y. O. Estrada, R. C. Genova, B. Girma, E. S. Kissel, A. N. Levy, S. MacCraken, P. R. Mastrandrea & L. E. White (Eds.). *Contribution of Working Group II to the Fifth Assessment Report of the Intergovernmental Panel on Climate Change* (pp. 1–113). Cambridge, UK: Cambridge University Press.

IPES. (2017). *Too big to feed: Exploring the impacts of mega-mergers, consolidation and concentration of power in the agri-food sector*. Retrieved from www.ipes-food.org/images/Reports/Concentration_FullReport.pdf

James, S. W. (2016). Beyond 'local' food: How supermarkets and consumer choice affect the economic viability of small-scale family farms in Sydney, Australia. *Area*, *48*(1), 103–110. doi:10.1111/area.12243

Jansen, K. & Gupta, A. (2009). Anticipating the future: 'Biotechnology for the poor' as unrealised promise? *Futures*, *41*, 436–445.

Jarroud, M. (2014, January 4). *Women farmers in Chile to teach the region agroecology*. Inter Press Service. Retrieved from www.ipsnews.net/2014/01/women-farmers-chile-teach-region-agroecology/

Johnston, J. & Baumann, S. (2010). *Foodies: Democracy and distinction in the gourmet foodscape*. Abingdon, Oxon: Routledge.

Johnston, J., Biro, A. & MacKendrick, N. (2009). Lost in the supermarket: The corporate-organic foodscape and the struggle for food democracy. *Antipode*, *41*(3), 509–532.

Jones, P. (2009). *Executive directive on healthy and sustainable food 09–03. San Francisco: City and County of San Francisco*. Retrieved from https://sfgov.org/sffood/sites/default/files/Documents/sffood/FPC_Summary_Report_December_2010_final1.pdf

Juris, J. (2008). *Networking futures: The movements against corporate globalisation*. Durham, NC: Duke University Press.

Kavada, A. (2013). Internet cultures and protest movements: The cultural links between strategy, organizing and online communication. In B. Cammaerts, A. Mattoni & P. McCurdy (Eds.), *Mediation and protest movements* (pp. 75–93). Chicago: University of Chicago Press.

Keck, M. & Sikkink, K. (1998). *Activists beyond borders: Transnational advocacy networks in international politics*. Ithaca: Cornell University Press.

Kenny, T. & Sage, C. (2018). Food surplus as charitable provision: Obstacles to reintroducing food as the commons. In J. L. Vivero-Pol, T. Ferrando, O. D. Schutter & U. Mattei (Eds.), *Routledge handbook of good as a commons*. Abingdon: Routledge.

Kerssen, T. M. & Brent, Z. W. (2017). Ground the US movement: Bringing land into food justice. In A. H. Alkon & J. Guthman (Eds.), *The new food activism: Opposition, cooperation and collective action* (pp. 284–315). Oakland, CA: University of California Press.

Kjaernes, U., Harvey, M. & Warde, A. (2007). *Trust in food*. New York: Macmillan.

Klein, N. (2014). *This changes everything: Capitalism versus the climate*. New York: Simon & Schuster.

Kneen, B. (2011). *Invisible giants: Cargill and its transnational strategies*. New York: Macmillan.

Knox, M. (2015). *Supermarket monsters: The price of Coles and Woolworths' dominance*. Carlton, VIC: Redback.

Koschmann, M. (2011). Developing a communicative theory of the nonprofit. *Management Communication Quarterly*, *26*(1), 139–146. doi:10.1177/0893318911423640

Kröger, M. (2011). Promotion of contentious agency as a rewarding movement strategy: Evidence from the MST-paper industry conflicts in Brazil. *Journal of Peasant Studies*, *38*(2), 435–458. doi:10.1080/03066150.2011.559016

Kucinich, E. (2017, December 6). Monsanto: The enemy of family farmers. *Huffpost*.

Kuhn, T. (2012). Negotiating the micro-macro divide: Thought leadership from organizational communication for theorizing organization. *Management Communication Quarterly*, *26*(4), 543–584. doi:10.1177/0893318912462004

La Vía Campesina. (2003). *What is food sovereignty?* Retrieved from the La Vía Campesina website https://viacampesina.org/en/food-sovereignty/

La Vía Campesina. (2008). *Vía Campesina proposal to solve the food crisis: Strengthening peasant and farmer-based food production*. Retrieved from www.foodfirst.org/en/node/2109

La Vía Campesina. (2009). *La Vía Campesina policy documents: 5th Conference*. Retrieved from https://viacampesina.org/en/wp-content/uploads/sites/2/2010/03/BOOKLET-EN-FINAL-min.pdf

La Vía Campesina. (2012). *A call to reject false capitalist solutions to climate crisis: World's peasant farmers help cool a warming planet*. Retrieved from www.commondreams.org/views/2012/12/07/call-reject-false-capitalist-solutions-climate-crisis

La Vía Campesina. (2013a). *Agroecology as a way of life*. Retrieved from the La Vía Campesina website https://viacampesina.org/en/agroecology-as-a-way-of-life/

La Vía Campesina. (2013b). *Vía Campesina TV*. Website https://tv.viacampesina.org/-ABOUT-?lang=en

La Vía Campesina. (2015). *Peasant agroecology cools the planet! Zero budget natural farming and the road to resistance at COP 21*. Retrieved from the La Vía Campesina website https://viacampesina.org/en/peasant-agroecology-cools-the-planet-zero-budget-natural-farming-and-the-road-to-resistance-at-cop-21/

La Vía Campesina (2017a). *Euskal Herria declaration*. Retrieved from the La Vía Campesina website https://viacampesina.org/en/viith-international-conference-la-via-campesina-euskal-herria-declaration/

La Vía Campesina. (2017b). *Agroecology, a way of life, struggle, and resistance against capitalism!* Retrieved from https://viacampesina.org/en/agroecology-way-life-struggle-resistance-capitalism/LVC

La Vía Campesina. (2017c). *Struggles of La Vía Campesina for agrarian reform and the defense of life, land and territories*. Retrieved from https://viacampesina.org/en/wp-content/uploads/sites/2/2017/10/compressed_Publication-of-Agrarian-Reform-EN.pdf

Lambie-Mumford, H. & Green, M. A. (2017). Austerity, welfare reform and the rising use of food banks by children in England and Wales. *Area*, *49*(3), 273–279.

Lamphere, J. A. & East, E. A. (2017). Monsanto's biotechnology politics: Discourses of legitimation. *Environmental Communication*, *11*(1), 75–89.

Lang, T. (2003). Food industrialisation and food power: Implications for food governance. *Development Policy Review*, *21*(5–6), 555–568.

Lang, T. (2005). Food control or food democracy? Re-engaging nutrition with society and the environment. *Public Health Nutrition*, *8*, 730–737.

Lang, T. & Heasman, M. A. (2004). *Food wars: The global battle for mouths, minds and markets*. London, UK: Earthscan.

Laughton, R. (2017). *A matter of scale: A study of the productivity, financial viability and multifunctional benefits of small farms (20ha and less)*. Retrieved from the La Vía Campesina website https://viacampesina.org/en/survey-shows-small-scale-agroecological-farms-produce-high-yields/

Lave, J. & Wenger, E. (1991). *Situated learning: Legitimate peripheral participation*. Cambridge: Cambridge University Press.

Lawrence, G., Richards, C. & Lyons, K. (2013). Food security in Australia in an era of neoliberalism, productivism and climate change. *Journal of Rural Studies*, *29*, 3039.

Lear, J. (2006). *Radical hope: Ethics in the face of cultural devastation*. Boston, MA: Harvard University Press.

Ledford, H. (2013). Seed-patent case in Supreme Court: Loss of patent control could rekindle 'terminator' technology. *Nature*, *494*, 7437.

Lee, C. W. (2015). Participatory practices in organizations. *Sociology Compass*, *9*(4), 272–288.

Lewis, T. (2018). Food politics in a digital era. In T. Schneider, K. Eli, C. Dolan & S. Ulijaszek (Eds.), *Digital food activism*. Abingdon: Routledge.

Loh, P. & Agyeman, J. (2017). Boston's emerging food solidarity economy. In A. H. Alkon & J. Guthman (Eds.), *The new food activism: Opposition, cooperation and collective action* (pp. 257–283). Oakland, CA: University of California Press.

London Food Charter. (2011). Retrieved from http://council.london.ca/meetings/CNC%20Reports/2011-03-29%20Report/Item%2010.pdf

London Urban Agricultural Strategy. (2017). Retrieved from www.london.ca/business/Planning-Development/current-topics/Documents/2017-Nov-UAS-Final.pdf

Lupton, D. (2018). Cooking, eating, uploading: Digital food cultures. In K. LeBesco & P. Naccarato (Eds.), *The handbook of food and popular culture*. London, UK: Bloomsbury.

Lupton, D. & Feldman, Z. (forthcoming). *Digital food cultures*. London: Routledge.

Machin, D. & Mayr, A. (2012). *How to do critical discourse analysis*. Los Angeles, CA: Sage.

Mackendrick, N. (2014). Foodscape. *Key Concepts in Social Research*, *13*(3), 16–18.

Madureira, A. M. & Baeten, G. (2016). By invitation only: Uses and users of the 'entrepreneurial city'. *International Planning Studies*, *21*(4), 362–376. doi:10.1080/13563475.2016.1157015

Mann, A. (2011). Lobbying for peasant rights at the UN. *Right to Food and Nutrition Watch*, 18–22.

Mann, A. (2013). Food sovereignty in a post-organic era. In B. Mascitelli & A. Lobo (Eds.), *Organics in the global food chain* (pp. 201–229). Ballarat: Connor Court Publishing.

Mann, A. (2014a). *Global activism in food politics: Power shift*. Basingstoke: Palgrave Macmillan.

Mann, A. (2014b). *Serving up food justice on the streets of New York*. Retrieved from the Sydney Environment Institute website http://sydney.edu.au/environment-institute/blog/serving-up-food-justice-on-the-streets-of-new-york/

Mann, A. (2016). *The right to food – and how 1.2 million Australians miss out*. Retrieved from The University of Sydney website https://sydney.edu.au/news-opinion/news/2016/04/11/-the-right-to-food--and-how-1-2-million-australians-miss-out.html

Mann, A. (2017a). Food sovereignty and the politics of food scarcity. In M. C. Dawson, C. Rosin & N. Wald (Eds.), *Global resources scarcity: Catalyst for conflict or cooperation?* (pp. 131–145). London, UK: Routledge.

Mann, A. (2017b). *Growing food in a post-truth era*. Retrieved from https://theconversation.com/growing-food-in-the-post-truth-era-78256

Mann, A. (2018a). Hashtag activism and the right to food in Australia. In T. Schneider, K. Eli, C. Dolan & S. Ulijaszek (Eds.), *Digital food activism* (pp. 168–184). Abingdon: Routledge.

Mann, A. (2018b). Food sovereignty: Deep histories, digital activism and the emergence of a transnational public. In M. Phillipov & K. Kirkwood (Eds.), *Alternative food politics: From the margins to the mainstream*. Oxfordshire, UK: Routledge.

Mann, A. (2018c). The politics of place: Networking resistance to coal seam gas mining. In B. Brevini & J. Lewis (Eds.), *Climate change and the media*. New York: Peter Lang.

Manzini, E. (2011). The new way of the future: Small, local, open and connected. *Social Space*, 99–105.

Martínez-Torres, M. E. & Rosset, P. (2014). Diálogo de saberes in La Vía Campesina: Food Sovereignty and Agroecology. *Journal of Peasant Studies*, *41*(6), 979–997.

Massey, C. (2018). *Call of the reed warbler: A new agriculture, a new earth*. Australia: University of Queensland Press.

Mattelart, A. (2000). *Networking the world: 1794–2000*. Minneapolis, MA: University of Minnesota Press.

Mayes, C. (2018). *Unsettling food politics: Agriculture, dispossession and sovereignty in Australia*. London, UK: Rowman & Littlefield.

McFarlane, C. (2011a). The city as a machine for learning. *Transactions of the Institute of British Geographers*, *36*, 360–376.

McFarlane, C. (2011b). *Learning the city: Knowledge and trans-local assemblage*. Chichester: Wiley-Blackwell.

McGuirk, J. (2015). *Radical cities: Across Latin America in search of a new architecture*. London: Verso.

McKeon, N. (2015). *Food security governance: Empowering communities, regulating corporations*. New York: Routledge.

McMichael, P. (2005). Global development and the corporate food regime. In F. H. Buttel & P. McMichael (Eds.), *New directions in the sociology of global development* (pp. 265–299). Bingley, UK: Emerald.

McMichael, P. (2008). Peasants make their own history, but not just as they please. *Journal of Agrarian Change*, *8*(2–3), 205–228. doi:10.1111/j.1471-0366.2008.00168.x

McMichael, P. (2009). A food regime genealogy. *Journal of Peasant Studies*, *36*(1), 139–169.

McMichael, P. (2013). *Food regimes and agrarian questions*. Black Point, Nova Scotia: Fernwood Publishing.

McMichael, P. (2014). Historicizing food sovereignty. *Journal of Peasant Studies*, *41*(6), 933–957.

McPhee, R. & Iverson, J. (2009). Agents of constitution in communidad: Constitutive processes of communication in organisations. In L. L. Putnam & A. M. Nicotera (Eds.), *Building theories of organisation: The constitutive role of communication* (pp. 49–87). New York, NY: Routledge.

McPhee, R. D. & Zaug, P. (2000). The communicative constitution of organisations. *The Electronic Journal of Communication*, *10*(1–2).

Meek, D. (2014). Agroecology and radical grassroots movements evolving moral economies. *Environment and Society: Advances in Research*, *5*, 47–65.

Meek, D. (2016). The cultural politics of the agricultural transition. *Agriculture and Human Values*, *33*, 275–290.

Meek, D. & Tarlau, R. (2016). Critical food systems education (CFSE): Educating for food sovereignty. *Agroecology and Sustainable Food Systems*, *40*(3), 237–260.

Meenar, M. & Hoover, B. (2012). Community food security via urban agriculture: Understanding people, place, economy and accessibility from a food justice perspective. *Journal of Agriculture, Food Systems, and Community Development*, *3*(1), 143–160.

Meenar, M., Morales, A. & Bonarek, L. (2017). Regulatory practices of urban agriculture: A connection to planning and policy. *Journal of the American Planning Association*, *83*(4), 389–403. doi:10.1080/01944363.2017.1369359

Meerow, S., Newell, J. P. & Stults, M. (2016). Defining urban resilience: A review. *Landscape and Urban Planning*, *147*, 38–49.

Mehta, L. (2011). The social construction of scarcity: The case of water in Western India. In R. Peet, P. Robbins & M. J. Watts (Eds.), *Global political ecology* (pp. 371–386). Abingdon, Oxfordshire: Routledge.

Mendes, W., Balmer, K., Kaethler, T. & Rhoads, A. (2008). Using land inventories to plan for urban agriculture: Experiences from Portland and Vancouver. *Journal of the American Planning Association*, *74*(4), 435–449.

Méndez, V. E., Bacon, C. M. & Cohen, R. (2013). Agroecology as a transdisciplinary, participatory, and action-oriented approach. *Agroecology and Sustainable Food Systems*, *37*(1), 3–18.

Micheletti, M. (2003). *Political virtue and shopping: Individuals, consumerism, and collective action*. New York: Palgrave Macmillan.

Milan Urban Food Policy Pact. (2015). Retrieved from Milan www.milanurbanfoodpolicypact.org/

Milner, N. (2017). The right to food is nature too: Food justice and everyday environmental expertise in the Salvadoran permaculture movement. *Local Environment*, *22*(6), 764–783.

Minaker, L. M., Fisher, P. F., Raine, K. R. & Lawrence, F. L. (2011). Measuring the food environment: From theory to planning practice. *Journal of Agriculture, Food Systems, and Community Development*, *2*(1), 65–82.

Minkoff-Zern, L. (2017). Farmworker-led food movements then and now: United farm workers, the coalition of Immokalee workers, and the potential for farm labor justice. In A. H. Alkon & J. Guthman (Eds.), *The new food activism: Opposition, cooperation and collective action* (pp. 157–178). Berkeley, CA: University of California Press.

Minkoff, D. C. & McCarthy, J. D. (2005). Reinvigorating the study of organisational processes in social movements. *Mobilisation: An International Journal*, *10*(2), 289–308.

Monsanto Company. (1979). *Without chemicals, life itself would be impossible* [advertisement]. *National Geographic*. Retrieved from www.flickr.com/photos/jbcurio/8740859605

Monsanto Company. (1995a). *Speeches: Robert B. Shapiro*. Retrieved from https://web.archive.org/web/19961111115403/www.monsanto.com/MonPub/InTheNews/Speeches/951028Shapiro_Robert.html

Monsanto Company. (1995b). *Speeches: Nicholas L. Reding*. Retrieved from https://web.archive.org/web/19961111121358/www.monsanto.com/MonPub/InTheNews/Speeches/951115Reding_Nicholas.html

Monsanto Company. (1996). *Monsanto values*. Retrieved from https://web.archive.org/web/19970714010223/www.monsanto.com/MonPub/NewMonsanto/Values/index.html

Monsanto Company. (2006). *Our pledge: Human rights policy*. Retrieved from https://web.archive.org/web/20061128202429/www.monsanto.com/monsanto/layout/our_pledge/human_rights/human_rights.asp?level1=WhoWeAre&level2=OurPledge

Monsanto Company. (2016). *Honey bee health*. Retrieved from the Monsanto website https://monsanto.com/company/sustainability/biodiversity/articles/bee-health/

Monsanto Company. (2017). Retrieved from website https://monsanto.com/

Monsanto Company. (2018a). *Digital Farming Technology: Protecting Crops Around the World*. Retrieved from https://monsanto.com/innovations/modern-agriculture/articles/digital-farming-technology-around-world/

Monsanto Company. (2018b). *Monsanto Spotlights Research & Development Pipeline Projects to Help Farmers Combat Threats, Optimise Opportunities and Conserve Natural Resources*. Retrieved from https://monsanto.com/news-releases/monsanto-spotlights-research-development-pipeline-projects-to-help-farmers-combat-threats-optimize-opportunities-and-conserve-natural-resources/

Monsanto Magazine. (1962, October). *The Desolate Year*, 4–9. Retrieved from https://iseethics.files.wordpress.com/2011/12/monsanto-magazine-1962-the-desolate-yeart.pdf

Moore, K. & Swisher, M. E. (2015). The food movement: Growing white privilege, diversity, or empowerment? *Journal of Agriculture, Food Systems, and Community Development*, *5*(54), 115–119. doi:10.5304/jafscd.2015.054.013

Moore-Lappe, F. & Du Bois, P. M. (1994). *The quickening of America: Rebuilding our nation, remaking our lives*. San Francisco: Jossey-Bass.

Morgan, K. (2015). Nourishing the city: The rise of the urban food question in the Global North. *Urban Studies*, *52*(8), 1379–1394.
Mosco, V. (1996). *The political economy tradition of communication: Rethinking and renewal.* London: Sage.
Mosco, V. (2008). Current trends in the political economy of communication. *Global Media Journal – Canadian Edition*, *1*(1), 45–63.
Mueller, M. L., Kuerbis, B. N. & Pagé, C. (2007). Democratising global communication? Global civil society and the campaign for communication rights in the information society. *International Journal of Communication*, *1*, 267–296.
Munck, R. (2010). Globalization, crisis and social transformation: A view from the South. *Globalizations*, *7*(1–2), 235–246.
Murdock, G. (1999). Rights and representations: Public discourse and cultural citizenship. In J. Gripsrud (Ed.), *Television and common knowledge* (pp. 7–17). London, UK: Routledge.
Murdock, G. & Golding, P. (2005). Culture, communication and political economy. In J. Curran & M. Gurevitch (Eds.), *Mass media and society* (4 ed., pp. 60–83). London, UK: Hodder Arnold.
National Rural Health Alliance. (2016). *Food insecurity and health in rural and remote Australia.* Retrieved from www.agrifutures.com.au/wp-content/uploads/publications/16-053.pdf
NCD Risk Factor Collaboration. (2016). Trends in adult body-mass index in 200 countries from 1975 to 2014: A pooled analysis of 1698 population-based measurement studies with 19.2 million participants. *Lancet*, *387*, 1377–1396.
Neslen, A. (2017, November 27). Controversial glyphosate weedkiller wins new five-year lease in Europe. *Guardian.* Retrieved from www.theguardian.com/environment/2017/nov/27/controversial-glyphosate-weedkiller-wins-new-five-year-lease-in-europe
Noddings, N. (1992). *The challenges to care in schools: An alternative approach to education.* New York: Teachers College Press.
Nordahl, D. (2014). *Public produce.* Washington DC: Island Press.
O'Cádiz, M. D. P., Wong, P. L. & Torres, C. A. (1998). *Education and democracy: Paulo Freire, social movements and educational reform in São Paulo.* Boulder, CO: Westview Press.
OHCHR. (2018). *Human rights council adopts five resolutions, including on Burundi, Syria, and on the rights of peasants.* Retrieved from the Office of the High Commissioner of United National Human Rights website www.ohchr.org/EN/HRBodies/HRC/Pages/NewsDetail.aspx?NewsID=23660&LangID=E
O'Neill, P. & Dimeski, B. (2013). *The origins and destinations of produce flows through Sydney Markets.* Retrieved from www.uws.edu.au/__data/assets/pdf_file/0010/482518/Sydney_Markets_Report.pdf
Ostrom, E. (1990). *Governing the commons: The evolution of institutions for collective action.* Cambridge: Cambridge University Press.
P. M. (2014). *The power of neighborhood and the commons.* New York: Autonomedia.
Paget-Clarke, N. (2014). Interview with Francisca Rodriguez. *In Motion Magazine.* Retrieved from www.inmotionmagazine.com/global/f-rodriguez-interview/francisca-rodriguez-intview_eng2.html#Anchor-space-51540
Participatory City Foundation. (2018). *Made to measure.* Retrieved from the Participatory City website www.participatorycity.org/made-to-measure-1

Pascoe, B. (2014). *Dark emu: Black seeds: Agriculture or accident?* Broome, WA: Magabala Books.

Patel, R. (2009a). *Stuffed and starved: Markets, power and the hidden battle for the world food system.* Melbourne: Black Inc.

Patel, R. (2009b). What does food sovereignty look like? *Journal of Peasant Studies, 36*(3), 587–612.

Patel, R. (2009c). Food sovereignty. *Journal of Peasant Studies, 36*(3), 663–706.

Patel, R. & Moore, J. W. (2017). *A history of the world in seven cheap things: A guide to capitalism, nature, and the future of the planet.* Oakland, CA: University of California Press.

Patel, R., Torres, R. G. & Rosset, P. (2005). Genetic engineering in agriculture and corporate engineering in public debate: Risk, public relations, and public debate over genetically modified crops. *International Journal of Occupational and Environmental Health, 11*(4), 428–436.

Peet, R., Robbins, P. & Watts, M. (Eds.). (2011). *Global political ecology* (1st edn). Abingdon, Oxfordshire: Routledge.

Peñaloza, F. & Walsh, S. (Eds.). (2019). *Mapping South-South connections: Australia and Latin America.* Basingstoke: Palgrave Macmillan.

Perfecto, I., Wright, A. & Vandermeer, J. H. (2009). *Nature's matrix: Linking agriculture, conservation and food sovereignty.* London: Routledge/Earthscan.

Phillips, T. (2018, October 22). Brazil's Jair Bolsonaro threatens purge of leftwing 'outlaws'. *Guardian.* Retrieved from www.theguardian.com/world/2018/oct/22/brazils-jair-bolsonaro-says-he-would-put-army-on-streets-to-fight

Pickett, S. T. A., Cadenasso, M. L. & Grove, J. M. (2004). Resilient cities: Meaning, models, and metaphor for integrating the ecological, socio-economic, and planning realms. *Landscape and Urban Planning, 69*(4), 369–384.

Poletta, F. (2002). *Freedom is an endless meeting: Democracy in American social movements.* Chicago: Chicago University Press.

Pollan, M. (2008). *In defence of food.* New York: Penguin.

Pothukuchi, K. (2004). Community food assessment A first step in planning for community food security. *Journal of Planning Education and Research, 23*(4), 356–377.

Pothukuchi, K. (2017). 'To allow farming is to give up on the city': Political anxieties related to the disposition of vacant land for urban agriculture in Detroit. *Journal of Urban Affairs, 39*(8), 1169–1189.

Pothukuchi, K. & Kaufman, J. L. (1999). Placing the food system on the urban agenda: The role of municipal institutions in food systems planning. *Agriculture and Human Values, 16*(2), 213–224.

PPS. (2008). *A guide to neighborhood placemaking in Chicago.* Retrieved from www.placemakingchicago.com/cmsfiles/placemaking_guide.pdf

Prody, J. (2013). A call for polycultural arguments: Critiquing the monoculture rhetoric of the local food movement. *Argumentation and Advocacy, 50*(2), 104–119.

Purcell, M. & Brown, J. C. (2005). Against the local trap: Scale and the study of environment and development. *Progress in Development Studies, 5*(4), 279–297.

Putnam, L. L. & Nicotera, A. M. (2009). *Building theories of organisation: The constitutive role of communication.* New York, NY: Routledge.

Radcliffe, S. (2002). Indigenous women, rights and the nation-state in the Andes. In N. Craske & M. Molyneux (Eds.), *Gender and the politics of rights and democracy in Latin America* (pp. 149–172). Basingstoke: Palgrave.

Raja, S., Born, B. M. & Russell, J. K. (2008). *A planners guide to community and regional food planning: Transforming food environments, facilitating healthy eating*. Retrieved from www.planningclimatechange.org/public/file/Planners%20Guide%20to%20Community%20and%20Regional%20Food%20Planning.pdf

Reitan, R. (2007). *Global activism*. New York: Routledge.

Renting, H., Schermer, M. & Rossi, A. (2012). Building food democracy: Exploring civic food networks and newly emerging forms of food citizenship. *International Journal of Sociology of Agriculture and Food*, *19*(3), 289–307.

Reuters. (2017, September 27). *Post office workers grow vegetables, breed chickens on Paris rooftop 'farm'*. Retrieved from https://uk.reuters.com/article/us-france-agriculture-urban-rooftop/post-office-workers-grow-vegetables-breed-chickens-on-paris-rooftop-farm-idUKKCN1C11UX

Reuters. (2018, August 23). *Bayer's Monsanto faces 8,000 lawsuits on glyphosate*. Retrieved from www.reuters.com/article/us-bayer-glyphosate-lawsuits/bayers-monsanto-faces-8000-lawsuits-on-glyphosate-idUSKCN1L81J0

Reynolds, K. & Cohen, N. (2016). *Beyond the kale: Urban agriculture and social justice activism in New York city*. Athens, Georgia: The University of Georgia Press.

Richard, A. (2012). 'Sin maíz no hay país': Citizenship and environment in Mexico's food sovereignty movement. In A. Latta & H. Wittman (Eds.), *Environment and citizenship in Latin America: Natures, subjects and struggles*. Amsterdam: Centre of Latin American Research and Documentation (CEDLA).

Richards, P. (2004). *Pobladoras, indígenas, and the state: Conflicts over women's rights in Chile*. New Brunswick: Rutgers University Press.

Riches, G. (2018). *Food bank nation: Poverty, corporate charity and the right to food*. London: Routledge.

Ricoeur, P. (1980). Narrative time. In W. Mitchell (Ed.), *On narrative*. Chicago: Chicago University Press.

Ricoeur, P. (1992). *Oneself as another*. Chicago: Chicago University Press.

Robin, M.-M. (2008). *The world according to Monsanto. Pollution, corruption, and the control of our food supply: An investigation into the world's most controversial company*. New York: The New Press.

Rochon, T. (1998). *Culture moves: Ideas, activism and changing values*. Princeton: Princeton University Press.

Rogers, E. M. (2004). *Diffusion of innovations*, 5th edition. New York: Free Press.

Rooke, D. (2018). *One last spin: The power and peril of the pokies*. Sydney: Scribe.

Rosier, K. (2011). *Food insecurity in Australia: What is it, who experiences it and how can child and family services support families experiencing it?* Retrieved from https://aifs.gov.au/cfca/publications/food-insecurity-australia-what-it-who-experiences-it-and-how-can-child

Rosset, P., Sosa, B. M., Jaime, A. M. R. & Lozano, D. R. A. (2011). The campesino-to-campesino agroecology movement of ANAP in Cuba: Social process methodology in the construction of sustainable peasant agriculture and food sovereignty. *Journal of Peasant Studies*, *38*(1), 161–191. doi:10.1080/03066150.2010.538584

Rumetshofer, L. & Kay, S. (2016). Youth and agriculture: Join the European food sovereignty movement! *Farming Matters*, *32*(1), 17.

Rundgren, G. (2016). Food: from commodity to commons. *Journal of Agricultural and Environmental Ethics*, *29*, 103–121.

Sage, C. (2014). The transition movement and food sovereignty: from local resilience to global engagement in food system transformation. *Journal of Consumer Culture*, *14*(2), 254–275.

Salazar, J. F. (2015, July 24). Buen Vivir: South America's rethinking of the future we want. *The Conversation*.

Salter, L. (2003). Democracy, new social movements, and the Internet: A Habermasian analysis. In M. McCaughey & M. D. Ayers (Eds.), *Cyberactivism. Online activism in theory and practice*. London: Routledge.

Samuels, G. (2018). *Independent review of the food and grocery code of conduct*. Retrieved from https://static.treasury.gov.au/uploads/sites/1/2018/10/Independent-review-of-the-Food-and-Grocery-Code-of-Conduct-Final-Report.pdf

Santos, B. de S. (Ed.) (2007). *Another knowledge is possible: Beyond northern epistemologies*. London: Verso.

Sanyé-Mengual, E., Anguelovski, I., Oliver-Solà, J., Montero, J. I. & Rieradevall, J. (2016). Resolving differing stakeholder perceptions of urban rooftop farming in Mediterranean cities: Promoting food production as a driver for innovative forms of urban agriculture. *Agriculture and Human Values*, *33*(1), 101–120.

Saragih, H. (2012). *Our time has come! Food sovereignty now!* Retrieved from http://viacampesina.org/en/index.php/main-issues-mainmenu-27/food-sovereignty-and-trade-mainmenu-38/1320-our-time-has-come-food-sovereignty-now

Savin-Baden, M. & Howell Major, C. (2012). *Qualitative research: The essential guide to theory and practice*, 1st edition. London: Routledge.

Sbicca, J. (2018). *Food Justice Now!: Deepening the roots of social struggle*. Minneapolis, MA: University of Minnesota Press.

Scharf, K., Levkoe, C. & Saul, N. (2010). *In every community a place for food: The role of the community food centre in building a local, sustainable, and just food system*. Retrieved from https://metcalffoundation.com/wp-content/uploads/2011/05/in-every-community.pdf

Schechinger, A. W. & Cox, C. (2016). *Feeding the world: Think US agriculture will end world hunger? Think again*. Retrieved from www.ewg.org/agmag/2016/10/think-us-agriculture-will-end-world-hunger-think-again

Schiavoni, C. M. (2015). Competing sovereignties, contested processes: Insights from the Venezuelan food sovereignty experiment. *Globalizations*, *12*(4), 466–480.

Schipper, E. L. F. (2009). Meeting at the crossroads? Exploring the linkages between climate change adaptation and disaster risk reduction. *Climate and Development*, *1*(1), 16–30.

Schluter, D. C. (2015). *Contested safety: Monsanto's 'Roundup ready' agricultural assemblage versus counter discourses of Roundup risk*. (Master's thesis), Arizona State University, United States of America. Retrieved from https://repository.asu.edu/attachments/150819/content/Schluter_asu_0010N_15102.pdf

Schneider, T., Eli, K., Dolan, C. & Ulijaszek, S. (Eds.). (2018). *Digital food activism*. Abingdon, Oxfordshire: Routledge.

Schortman, A. (2010). The children cry for Burger King: Modernity, development, and fast food consumption in northern Honduras. *Environmental Communication*, *4*(3), 318–337.

Schumacher, E. F. (2011). *Small is beautiful: A study of economics as if people mattered*. London: Vintage Books.

Scoones, I., Edelman, M., Borras, S. M., Hall, R., Wolford, W. & White, B. (2017). Emancipatory rural politics: Confronting authoritarian populism. *Journal of Peasant Studies*, *45*(1), 1–20. doi:10.1080/03066150.2017.1339693

Scott, J. (1985). *Weapons of the weak: Everyday forms of peasant resistance*. New Haven, CT: Yale University Press.

Sen, A. (2002). *Rationality and freedom*. Cambridge, MA: Harvard University Press.

Sen, N. J. (2017). *More rooftop gardens, urban farms planned*. Retrieved from www.straitstimes.com/singapore/environment/more-rooftop-gardens-urban-farms-planned

Sepulveda, P. (2009). Agricultural boom passes women farmers. *IPS News*. Retrieved from www.ipsnews.net/2009/03/chile-agricultural-boom-passes-women-farmers-by/

Seufert, P., Mendonca, M. L. & Pitta, F. (2018). When land becomes a global financial asset: The MATOPIBA case in Brazil. *Right to Food and Nutrition Watch: When Food Becomes Immaterial: Confronting the Digital Age*, *10*, 20–25.

Sharing Lab. (2017). *Let's stop talking about smart cities*. Retrieved from https://medium.com/we-research-and-expriment-with-how-the-sharing/lets-stop-talking-about-smart-cities-df5a0d187b51

Shaw, D. J. (2007). *World food security: A history since 1945*. New York: Palgrave Macmillan.

Sherwood, S., Deaconu, A. & Paredes, M. (2017a). 25,000 families campaign: The existence of flavour and taste. In S. Sherwood, A. Arce & M. Paredes (Eds.), *Food agriculture and social change: The everyday vitality of Latin America* (pp. 198–210). Abingdon, Oxford: Routledge.

Sherwood, S., Fisher, E. & Arce, A. (2017b). The vitality of everyday food. In S. Sherwood, A. Arce & M. Paredes (Eds.), *Food agriculture and social change: The everyday vitality of Latin America* (pp. 211–223). Abingdon, Oxford: Routledge.

Shiva, V. (1993). *Monocultures of the mind: Perspectives on biodiversity and biotechnology*. London: Zed Books.

Sikes, P. (2004). Methodology, procedures and ethical concerns. In C. Opie (Ed.), *Doing educational research: A guide for first time researchers*, 15–33. London: Sage.

Skeggs, B. (1997). *Formations of class and gender: Becoming respectable*. London: Sage.

Smith, J. (2005). Globalisation and social movement organisations. In G. Davis, D. McAdam, W. R. Scott & M. N. Zald (Eds.), *Social movements and organisation theory* (pp. 226–248). Cambridge: Cambridge University Press.

Sonnino, R. (2017). The cultural dynamics of food governance. City, culture and society. Retrieved from ScienceDirect website https://doi.org/10.1016/j.ccs.2017.11.001

Sosa, B. M., Jaime, A. M. R., Lozano, D. R. A. & Rosset, P. M. (2013). *Agroecological revolution: The farmer-to-farmer movement of the ANAP in Cuba*. Jakarta, Indonesia: La Vía Campesina.

Specht, K., Weith, T., Swoboda, K. & Siebert, R. (2016). Socially acceptable urban agriculture businesses. *Agronomy for Sustainable Development*, *36*(1).

Spurling, N., McMeekin, A., Shove, E., Southerton, D. & Welch, D. (2013). *Interventions in practice: Re-framing policy approaches to consumer behaviour*. Retrieved from www.sprg.ac.uk/uploads/sprg-report-sept-2013.pdf

Steel, C. (2013). *Hungry city: How food shapes our lives.* London: Random House.

Stock, P., Carolan, M. & Rosin, C. (2015). *Food utopias: Reimagining citizenship, ethics and community.* London: Routledge.

Strodthoff, I. (2014). *Chile and Australia: Contemporary transpacific connections from the South.* Basingstoke: Palgrave Macmillan.

Stulka, S., Moore, L., Eicher-Miller, H. & Franzen-Castle, L. (2018). Voices for food: Methodologies for implementing a multi-state community-based intervention in rural, high poverty communities. *BCM Public Health, 18*(1), 1055.

Sullivan, K. (2018, November 13). Foodbank funding cuts risk more than $8m worth of food for needy families. *ABC News.* Retrieved from www.abc.net.au/news/2018-11-12/foodbank-funding-to-feed-hungry-australians-almost-halved/10485182

Sussman, L. & Bassarab, K. (2016). *Food policy council report 2016.* Retrieved from https://assets.jhsph.edu/clf/mod_clfResource/doc/FPC%20Report%202016_Final.pdf

Tarlau, R. (2014). From a language to a theory of resistance: Critical pedagogy, the limits of 'framing', and 'social change'. *Education Theory, 64(*4), 369–392.

Taylor, J. R. (1993). *Rethinking the theory of organizational communication: How to read an organization.* Norwood: Ablex.

Taylor, J. R. (2009). Organising from the bottom up? Reflections on the constitution of organisation in communication. In L. L. Putnam & A. M. Nicotera (Eds.), *Building theories of organisation: The constitutive role of communication* (pp. 153–186). New York, NY: Routledge.

Taylor, M. (2015). *The political ecology of climate change adaptation: Livelihoods, agrarian change and the conflicts of development.* London: Routledge/Earthscan.

Third World Network. (1998). *The model law of the organisation of African unity on community rights and on the control of access to biological resources.* Retrieved from www.twn.my/title/oau-cn.htm

Thomas, J., Barraket, J., Wilson, C. K., Cook, K., Louie, Y. M. & Holcombe-James, I. with Ewing, S. & MacDonald, T. (2018). *Measuring Australia's digital divide: The Australian digital inclusion index 2018*, RMIT University, Melbourne, for Telstra. Retrieved from https://digitalinclusionindex.org.au/wp-content/uploads/2018/08/Australian-digital-inclusion-index-2018.pdf

Thompson, E. P. (1971). The moral economy of the English crowd in the eighteenth century. *Past and Present, 50*, 76–136.

Thompson, E., Harper, A. & Kraus, S. (2008). *Think globally ~ Eat locally: San Francisco Foodshed Assessment.* Retrieved from San Francisco https://4aa2dc132bb150caf1aa-7bb737f4349b47aa42dce777a72d5264.ssl.cf5.rackcdn.com/ThinkGloballyEatLocally-FinalReport8-23-08.pdf

Tilly, C. & Tarrow, S. (2007). *Contentious politics.* Boulder: Paradigm.

Tompkins, E. L., Adger, W. N., Boyd, E., Nicholson-Cole, S., Weatherhead, K. & Arnell, N. (2010). Observed adaptation to climate change: UK evidence of transition to a well-adapting society. *Global Environmental Change, 20*(4), 627–635.

Trinidad, A. M. O. (2012). Critical indigenous pedagogy of place: A framework to indigenize a youth food justice movement. *Journal of Indigenous Social Development, 1*(1), 1–17.

Tsing, A. L. (2005). *Friction: An ethnography of global connection.* Princeton, NJ: Princeton University Press.

Tsing, A. L. (2015). *The mushroom at the end of the world*. Princeton, NJ: Princeton University Press.

UNDESA. (2010). *World Urbanization Prospects: The 2009 Revision Highlights*. Retrieved from www.edocs.unep.org/handle/20.500.11822/18401?show=full

Valencia, R. R. (1997). *The evolution of deficit thinking: Educational thought and practice*. London: Routledge.

Valenzuela, H. (2016). Agroecology: A global paradigm to challenge mainstream industrial agriculture. *Horticulturae, 2*(1). doi:10:3390/horticulturae2010002

Vallauri, U. (2014). Transition Belsize veg bag scheme: The role of ICTs in enabling new voices and community alliances around local food production and consumption. *Futures*, *62*, 173–180.

Valley, W., Wittman, H., Jordan, N., Ahmed, S. & Galt, R. (2018). An emerging signature pedagogy for sustainable food systems education. *Renewable Agriculture and Food Systems*, *33*(5), 467–480. doi:10.1017/S1742170517 000199

Vallianatos, E. (2017). Why honeybees don't have a chance in the midst of pesticides. *Huffpost*.

van der Ploeg, J. (2008). *The new peasantries: Struggles for autonomy and sustainability in an era of empire and globalization*. Oxford: Earthscan.

van Dijk, T. A. (1998). *Ideology: A multidisciplinary approach*. London: Sage.

Vanolo, A. (2014). Smartmentality: The smart city as disciplinary strategy. *Urban Studies*, *51*(5), 883–898.

Veltmeyer, H. (2010). The global crisis and Latin America. *Globalizations*, *7*(1), 217–233.

Vidal, J. (2018). Who should feed the world: Real people or faceless multinationals? Retrieved from *Guardian* website www.theguardian.com/commentisfree/2018/jun/05/feed-the-world-real-people-faceless-multinationals-monsanto-bayer

Vidal, A., Venkatesh, S., Brown, P. & Chaskin, R. (2001). *Building community capacity*. London: Routledge.

Vinson, T. (2009). *Social inclusion: The origins, meaning, definitions and economic implications of the concept of inclusion/exclusion*. Canberra: Australian Government.

Vivero-Pol, J. L. (2017). Food as commons or commodity? Exploring the links between normative valuations and agency in food transition. *Sustainability*, *9*, 442.

Voigt, K. (2011). Pigs in the backyard or the barnyard: Removing zoning impediments to urban agriculture. *Boston College Environmental Affairs Law Review*, *38*(2), 537–589.

von Bülow, M. (2010). *Building transnational networks: Civil society and the politics of trade in the Americas*. New York, NY: Cambridge University Press.

Wade, R. (2009). From global imbalance to global reorganisations. *Cambridge Journal of Economics*, *33*, 539–562.

Wasko, J. (2005). Studying the political economy of media and information. *Comunicação & Sociedade*, *7*, 25–48.

Watts, J. (2018). Murdered land activist adds to rising death toll in Brazil's Amazon. Retrieved from *Guardian*, Australia edition, website www.theguardian.com/environment/2018/apr/17/murdered-indigenous-land-activist-adds-to-rising-death-toll-in-brazils-amazon

Weasel, L. (2008). *Food fray: Inside the controversy over genetically modified food*. New York: AMACOM.

Weber, K. & King, B. (2014). Social movement theory and organisational studies. In P. Adler, P. D. Gay, G. Morgan & M. Reeds (Eds.), *Oxford handbook of sociology, social theory and organisational studies* (pp. 487–509). Oxford: Oxford University Press.

Weis, A. (2007). *The global food economy: The battle for the future of farming*. London, UK: Zed Books.

Wenger, E. (1998). Communities of practice: Learning, meaning and identity. *Journal of Mathematics Teacher Education*, 6(2), 185–194.

West, C. (1993). *Keeping faith*. New York: Routledge.

White, J. M. (2015). Anticipatory logics of the smart city's global imaginary. *Urban Geography*, *37*(4), 572–589.

Williams, M. (2017). Searching for *actually existing* justice in the city. *Urban Studies*, *54*(10), 2217–2231.

Williams, R. (1961). *The long revolution*. Harmondsworth: Penguin.

Wolford, W, (2010). *This land is ours now: Social mobilisation and the meanings of land in Brazil*. Durham, NC: Duke University Press.

Wright, C. & Nyberg, D. (2015). *Climate change, capitalism and corporations: Processes of creative self-destruction*. Cambridge: Cambridge University Press.

Yashar, D. (2005). *Contesting citizenship in Latin America: The rise of indigenous movements and the post-liberal challenge*. New York, NY: Cambridge University Press.

Young, I. M. (1990). *Justice and the politics of difference*. Princeton, NJ: Princeton University Press.

Young, I. M. (1996). Communication and the other: Beyond deliberative democracy. In S. Benhabib (Ed.), *Democracy and difference: Contesting the boundaries of the political* (pp. 121–135). Princeton, NJ: Princeton University Press.

Young, I. M. (2000). *Inclusion and democracy*. Oxford: Oxford University Press.

Zacune, J. (2012). *Combatting Monsanto: Grassroots resistance to the corporate power of agribusiness in the era of the 'green economy' and a changing climate*. Retrieved from the La Vía Campesina website https://viacampesina.org/en/wp-content/uploads/sites/2/2012/04/Monsanto-Publication-EN-Final-Version.pdf

Zautra, A. J., Hall, J. S. & Murray, K. E. (2010). Resilience: A new definition of health for people and communities. In J. Reich, A. J. Zautra & J. S. Hall, (Eds.), *Handbook of adult resilience* (pp. 3–30). New York: The Guildford Press.

Zundel, T. & Ribeiro, S. (2018). Let them eat data. *Right to Food and Nutrition Watch: When Food Becomes Immaterial: Confronting the Digital Age*, *10*, 26–31.

100 Resilient Cities. (2019). Retrieved from http://100resilientcities.org/challenges/food-insecurity/

Index